P9-AZX-208

The
Search

The Search

How Google and Its Rivals Rewrote the Rules of Business and Transformed Our Culture

John Battelle

Portfolio

PORTFOLIO
Published by the Penguin Group
Penguin Group (USA) Inc., 375 Hudson Street,
New York, New York 10014, U.S.A.
Penguin Group (Canada), 90 Eglinton Avenue East, Suite 700,
Toronto, Ontario, Canada M4P 2Y3 (a division of Pearson Penguin Canada Inc.)
Penguin Books Ltd, 80 Strand, London WC2R 0RL, England
Penguin Ireland, 25 St. Stephen's Green, Dublin 2, Ireland (a division of Penguin Books Ltd)
Penguin Books Australia Ltd, 250 Camberwell Road, Camberwell,
Victoria 3124, Australia (a division of Pearson Australia Group Pty Ltd)
Penguin Books India Pvt Ltd, 11 Community Centre, Panchsheel Park,
New Delhi–110 017, India
Penguin Group (NZ), Cnr Airborne and Rosedale Roads, Albany,
Auckland 1310, New Zealand (a division of Pearson New Zealand Ltd)
Penguin Books (South Africa) (Pty) Ltd, 24 Sturdee Avenue,
Rosebank, Johannesburg 2196, South Africa

Penguin Books Ltd, Registered Offices:
80 Strand, London WC2R 0RL, England

First published in 2005 by Portfolio,
a member of Penguin Group (USA) Inc.

10 9 8 7 6 5 4 3 2 1

Library of Congress Cataloging-in-Publication Data
Battelle, John, 1965–
The Search : how Google and its rivals rewrote the rules of business and transformed our culture / John Battelle
p. cm.
Includes index.
Contents: The database of intentions—Who, what, where, why, when, and how (much)—Search before Google—Google is born—A billion dollars, one nickel at a time—Google 2000–2004: zero to $3 billion in five years—The search economy—Search, privacy, government, and evil—Google goes public—Google today, Google tomorrow—Perfect search.
ISBN 1-59184-088-0
1. Google (Firm) 2. Internet industry—United States. 3. Web search engines. 4. Google. 5. Internet searching. 6. Information society—United States. I. Title: Google and its rivals rewrote the rules of business and transformed our culture. II. Title.

HD9696.8.U64G633 2005
338.7'6102504'0973—dc22 2005047538

Printed in the United States of America
Set in Adobe Garamond with Catull Regular
Designed by Daniel Lagin

For Michelle

Contents

Chapter 1
The Database of Intentions

The library of Alexandria was the first time humanity attempted to bring the sum total of human knowledge together in one place at one time. Our latest attempt? Google.
—Brewster Kahle, entrepreneur and founder, the Internet Archive

Everyone their own Boswell.
—Geoffrey C. Bowker, Department of Communication, University of California, San Diego

By the fall of 2001, the Internet industry was in full retreat. Hundreds of once promising start-ups—mine among them—lay smoldering in bankruptcy. The dreams of Internet riches, of changing the world of business and reshaping our culture in the process, dreams celebrated in magazine cover stories and television specials and unheard-of stock market valuations, well, those dreams were stone-cold dead.

Still smarting from the loss of my own Internet business[1] and wondering whether the Internet story could ever pick itself up off the ground, I stumbled across a link to the first edition of Google Zeitgeist. Zeitgeist is a clever public relations tool that summarizes search terms that are gaining or losing momentum during a particular period of time. By watching and counting popular search terms, Zeitgeist provides a fascinating summary of what our culture is looking for or finds interesting, and, conversely, what was once popular that is losing cultural momentum.

Since 2001, Google has maintained a weekly Zeitgeist on its press relations site, but the link I found was the company's first-ever version of the tool, and it summarized the entire year.[2] And what a year 2001 was! Listed among the top gaining queries were Nostradamus (number one), CNN (number two), World Trade Center (number three), and anthrax (number five). The only term to break into the top five that was not related to the terrorist attacks? A collective fantasy about magic and children, *Harry Potter*, at number four.

The fastest-declining queries demonstrated how quickly our culture was abandoning frivolity: Pokemon was number one, followed closely by Napster, *Big Brother* (a reality television show), *X-Men,* and the woman who won *Who Wants to Marry a Multi-Millionaire.*

I was transfixed. Zeitgeist revealed to me that Google had more than its finger on the pulse of our culture, it was directly jacked into the culture's nervous system. This was my first glimpse into what I came to call the Database of Intentions—a living artifact of immense power. *My God,* I thought, *Google knows what our culture wants!* Given the millions upon millions of queries streaming into its servers each hour, it seemed to me that the company was sitting on a gold mine of information. Entire publishing businesses could be created from the traces of intent evident in such a database; in fact, Google had already started its first: a beta project called *Google News*. Could it not also start a research and marketing company capable of telling clients exactly what people were buying, looking to buy, or avoiding? How about starting an e-commerce firm that already knew what the buyer wanted? How about a travel business that knew where the customer wanted to go? The possibilities, it seemed, were endless.

Not to mention that within Google's rich database lay potential fieldwork for thousands of doctorates in cultural anthropology, psychology, history, and sociology. This little company, I thought to myself, rapt and a bit naively, is holding the world by the thoughts. I've got to go see it. Maybe the dot-com dream wasn't dead; perhaps it had simply been hiding behind the implacable facade of a Google search box.

I remembered that back in April 2001, Eric Schmidt, a founder of Sun Microsystems, had left his job running Novell, the perpetually struggling networking giant, and accepted the chairman and CEO role over at Google (the industry was baffled by the move, but we'll get to that story later). I knew Eric somewhat, as I covered Novell and Sun while I was a trade reporter, and ran into him at various conferences during my career as an editor and publisher. I decided to take a chance and shoot him an e-mail. I really had no idea what I wanted to talk about, other than my nascent sense that he was onto something big.[3] Google, it seemed, was thriving. I had heard that it was pretty much the only place left in the Valley that was hiring engineers. Eric agreed to a meeting, and in early 2002, we sat down for the first of several intriguing talks.

Eric Looks for the Billion-Dollar Opportunity

When we met, I hadn't yet figured out I wanted to write this book, but I was headed that way. I introduced my concept of the Database of Intentions and spoke of how Zeitgeist scratched the surface of what seemed to be a massive new wealth of cultural understanding. As we spoke, I outlined how Google might create a media division to tap into that resource. Yahoo had already declared itself a media company, so why not Google? While Eric agreed that the data collected by Google was impressive, he didn't see the point of starting a media business. Google was a technology business, he told me. Media is best left to people like you, he added.

I argued that the two were intertwingled at Google, that his newly installed revenue base, AdWords, was pure advertising dollars:

media, in other words. Google's future, I counseled, was to be a media company. Eric disagreed. "We're looking for the next billion-dollar market in technology," he said. "Got any ideas?"

I didn't, but I came away from that meeting convinced that sooner or later, Google would take its place as a giant in the media landscape. It didn't take long. A year later I met with Eric again. Among his first words: "Isn't the media business great?"

In essence, Google and its competitors have created the first application to leverage the Database of Intentions in a commercial manner: paid search. In less than five years, the business has grown from next to nothing to more than $4 billion in revenue, and it is predicted to quadruple in another five years.

Along the way, search has moved from a useful service on the edge of most Internet users' experience to the de facto interface for computing in the information age. "As the amount of information available to us explodes, search has become the user's interface metaphor," observes Raymie Stata, a Silicon Valley–based engineer and entrepreneur. "There is now all this information that is possible to get into your hands. Search is our attempt to make sense of it."

In the past few years, search has become a universally understood method of navigating our information universe: much as the Windows interface defined our interactions with the personal computer, search defines our interactions with the Internet. Put a search box in front of just about anybody, and he'll know what to do with it. And the aggregate of all those searches, it turns out, is knowable: it constitutes the database of our intentions.

Search as Material Culture

As with many in the technology industry, my fascination with computers started with the Macintosh. Back in the mid-1980s I was an undergraduate studying cultural anthropology, and I took a class that focused on the idea of material culture—basically, interpreting the artifacts of everyday life. Professor Jim Deetz, a genteel Maryland native who favored contemporary Kentucky bourbon and nineteenth-century Virginian architecture, taught that the tools of archaeology—usually applied only in the context of civilizations long dead—should be used to interpret the lessons of cultural anthropology, which focuses on living cultures.

Deetz encouraged us to see all things modified by humankind as material culture, even if they weren't material in the atomic sense. Most interestingly, he encouraged us to interpret communication—in particular language and its written counterpart—as reflecting the culture that created it, fraught with all kinds of intent, controversy, politics, and relationships. Nothing you wouldn't find in a college literature or philosophy course, but this was also a science. Viewing language as artifact was a way to pick up current culture and hold it in your hand, make sense of it, read it.

Around the same time I was making some folding money beta testing WYSIWYG (What You See Is What You Get) software on a brand-spanking-new Macintosh, vintage 1984. Like just about everyone who used a Macintosh in those early days, I was hooked on the seductive blend of interface and execution—I pointed *there* and things changed . . . *there*. Anthropology and technology merged, and I was soon convinced that the Macintosh represented humankind's most sophisticated and important artifact ever: a representation of the plastic mind made visible. (Yeah, college—*exhaaaaale*—wasn't it great?)

Anyway, the idea that a WYSIWYG graphical user interface—especially when networked to others—could provide a medium connecting human intelligence drove much of my fascination with reporting on computing technology as a cultural artifact. From *Wired* to *The Industry Standard,* the "Mac as the greatest artifact" meme became one of my standard conversational riffs. I'd use it to frame conversations with writers, pitches to venture capitalists, and joints-after-midnight arguments with good friends. While others argued that the wheel or the internal combustion engine was civilization's greatest tool, I'd stick to my guns and argue for the Mac.

But once I'd seen Google's Zeitgeist, I knew my beloved Macintosh had been trumped. Every day, millions upon millions of people lean forward into their computer screens and pour their wants, fears, and intentions into the simple colors and brilliant white background of Google.com. "Peugeot dealer Lyon," one might ask (in French, of course). "Record criminal Michael Evans," an anxious woman might query as she awaits her blind date. "Toxic EPA Westchester County," a potential homeowner might ask, speaking in the increasingly ubiquitous, sophisticated, and evolving grammar of the Google search keyword.

Of course, the same is true for the search boxes at Yahoo, MSN, AOL, Ask, and hundreds of other Internet search, information, and commerce sites. Billions of queries stream across the servers of these Internet services—the aggregate thoughtstream of humankind, online. What are we creating, intention by single intention, when we tell the world what we want?

Link by link, click by click, search is building possibly the most lasting, ponderous, and significant cultural artifact in the history of humankind: the Database of Intentions. The Database of Intentions is simply this: the aggregate results of every search ever entered, every result list ever tendered, and every path taken as a result. It lives in many places, but three or four places in particular—AOL, Google, MSN, Yahoo—hold a massive amount of this data. Taken together, this information represents a real-time history of post-Web culture—a massive clickstream database of desires, needs, wants, and preferences that can be discovered, subpoenaed, archived, tracked, and exploited for all sorts of ends.

Consider the Database of Intentions as rich data topsoil on an archaeological layering of technology that over the past half century or so has created the potential for an entirely new culture to emerge. It's easy to consider the Web a relatively recent development, but the Web itself is built on the Internet, which in turn is built on a vast network of computers of all stripes—mainframes, minicomputers, powerful servers, the desktop PC, and any number of mobile devices. This network has been built over nearly three generations, yet only in the past decade has it emerged in our cultural consciousness. In the next decade, it will expand to our televisions, our automobiles, and our public spaces—nearly everything that can have a chip in it will have a chip in it, and nearly everything with a chip will become a node in humanity's ever-growing Database of Intentions.

This structure will provide the seedbed for scores of new cultural phenomena over the next decade. We've already seen it flower with services like Yahoo, Napster, eBay, and Google. And we're just at the beginning: in 2003 and 2004, hundreds of new companies sporting innovative, search-based models emerged—from entirely new forms of expression like blogging to personalized photography sites like Flickr. And at its core, all of this new growth starts with one person in front of a screen, typing in a query.

But Why Search?

Why, nearly everyone I hold dear has asked me at one point or another, why are you writing a book about search? A book about Google as a business, sure, they could understand that (and don't ask me how many folks thought I should have timed it to Google's public offering). But a book about . . . search? Might as well write about e-mail or the browser; both are as ubiquitous—and as boring. If you want a real insider narrative, I've often been counseled, you should write about your experiences with *Wired* or *The Industry Standard,* or get Larry Page and Sergey Brin (Google's founders) to sit down with you for an authorized business biography. But I couldn't imagine more dreadful topics. Books have already been written on my two previous companies, and I've actually read them both—putting me in pretty rare company. And Larry and Sergey have been furtive quarry; they are wary of a tell-all book on a company that they believe, quite appropriately, is still a work in progress.

So why search? As Google's extraordinary cultural aura illustrates, search has about it a whiff of the mysterious and the holy. But most specifically, through search one can tell the story of the modern Internet era in all its cultural and commercial nuances—from its beginnings in the early 1990s to its myriad potential futures.

Through applications like Archie, Gopher, and others, search was one of the first useful services to inhabit the Internet (after all, what's the point of the Net if you can't find anything?). Later, search became one of the first applications to adopt an actual business model—that of banner advertising. And with the Netscape IPO of 1995, search (and its partner, the browser) fired the Internet bubble's starting pistol.

Search—or more aptly, Web traffic, search's first cousin—drove the late-1990s mania with all things Internet. And even though that bubble burst, search continued to prosper as an application and a business model—many investors may have gotten soaked, but Internet users never stopped searching. Companies like Overture and Google made their first profits in the darkest hours of the dot-com collapse.

And search is smack in the middle of the Web's second coming, a resurgence driven by companies like Google, eBay, Amazon, Yahoo, and Microsoft. These companies are in an all-out war for the market of the future, one where the spoils number in the hundreds of billions of dollars. That alone is a pretty damn good reason to learn more about search. But those are the easy answers. Search drove the Internet and continues to do so, and search has created Google, certainly one of the most intriguing and successful companies of the Internet age. But somehow the idea of writing a book that starred only Google seemed an act of premature composition—the story has a beginning and a middle, but as yet, no end.

So while this book has, as its core, the story of Google, I believe the idea of search is bigger than any one company, and the impact of search on our culture is extraordinarily far reaching. For example, besides its obvious role as the driver of the commercial Internet, search will be the application that finally catalyzes the fabled convergence of television and personal computer—what is a cable television program guide, after all, but a second-rate search application yearning to be free?

Search and the Man-Machine Interface

Search is also a catalyst in promising attempts at cracking one of mankind's most intractable problems: the creation of artificial intelligence. By its nature search is one of the most challenging and interesting problems in all of computer science, and many experts claim that continued research into its mysteries will provide the commercial and academic mojo to allow us to create computers capable of acting, by all measures, like a human being.

In short, search may well lead to the creation of Hal, the intelligent but creepy computer doppelgänger of Stanley Kubrick's *2001: A Space Odyssey*. Or, if that possibility doesn't keep you up at night, think of search as the application that lays the foundation for Skynet, the AI program that takes over the world as imagined in the *Terminator* films, or the equally dystopian *Matrix* trilogy. We are fascinated by the man-versus-machine narrative barn burner; it dominates our cultural landscape. And search is the most likely candidate to bring any of these possiblilities to fruition. Call me paranoid (at least I have good company) but that alone makes search worth understanding.

Search will also be the way we rewire the relationship between ourselves and our government—a significant claim, to be sure, but one that can be backed up. Before I take this concept too far, I must acknowledge the fact that as I've described it thus far, the Database of Intentions does not exist. John Poindexter's attempts notwithstanding,[4] there is no great database in the sky, tracking our every move online. Our clickstream—the exhaust of our online lives—is scattered across a vast landscape of Internet sites and private machines, for the most part uncollected, uncategorized, mute.

But that is changing, and quickly. Just ten years ago, bandwidth was scarce and storage was expensive. Use of the Internet was comparatively sparse, files were small, and Internet companies, for the most part, didn't keep their log files—storing that data was too expensive. In the past few years, a good portion of our digitally mediated behavior—be it in e-mail, search, or the relationships we have with others—has moved online.

Why? The average cost per megabyte for storage has plummeted, and it will continue to drop to the point where it essentially reaches zero. At the same time, bandwidth has increased dramatically, and with it, usage—the Internet is now a permanent fixture in the majority of American homes and businesses. In essence, we have taken much of our once-ephemeral and quotidian lives—our daily habits of whom we talk to, what we look for, what we buy—and made those actions eternal. It is as if each of us, every day, is tracing a picture of Joycean complexity—recording the mundane and extraordinary course of our lives—via our interactions with the Internet, be they through our personal computers, our telephones, or our music players, and our interactions with businesses, either online or in the store (after all, that grocery club card information has to go somewhere, right?).

Cast your mind back to the pre-Web days, the PC era of 1985–1995. In this phase of the computing revolution, we brought our habitual presumptions to the practice of communication and discovery via the computer keyboard. We assumed (rightly or wrongly) that there was no permanent record of our actions on the computer. When we rummaged through our hard drives or, later, across LANs and WANs, we assumed the digital footprint we left behind—our clickstream—was as ephemeral as a phone call. Why would it be anything but? Clickstreams had no value beyond the action they predicated, serving only as a means to an end of finding a file or passing along a message.

The same assumptions clothed our e-mail. Sure, we understood that e-mail might reside (briefly) on servers, but for years we assumed that they were *our* e-mails, and the ISP or network over which they passed had no right to examine or manipulate them, much less own them. (In fact, the Electronic Communications Privacy Act of 1986 codified this sentiment into law, at least for private e-mail.) While the more sophisticated e-mail user among us has grown to understand the folly of this assumption in a corporate environment, the idea that e-mail is an ephemeral medium is still widely held. In 2003, Frank Quattrone, one of the technology sector's most powerful bankers and hardly a computing rube, was brought down by such a presumption when incriminating e-mails were used as evidence against him in a widely publicized trial.

But for most of us, the possibility of such negative consequences is remote; we still believe e-mail is an intensely private and ephemeral form of communication. And this holds true even when that e-mail lives on the servers of yahoo.com, hotmail.com, or gmail.com.

Finally, back in the PC era, the very idea that our relationships with others (our social network) or our relationships to goods and services (our commercial network) were anything but ephemeral was presumed: without the Internet, how could it be otherwise? Sure, once in a long while someone got a hold of your calling card, your little black book, or your credit card slip, and your privacy and security were breached, but as with e-mail, the chances of this occurring were so minute as to be irrelevant. Before the rise of Internet-based social networking services like Linked In or Friendster, social networks were simply records in your private contact database.[5]

In short, before the Web, we could pretty safely assume that our digitally mediated habits—rummaging through our hard drives, checking our e-mail, or looking up our contacts—were ephemeral, known only to us (and soon forgotten by us, to boot).

But now, details of our lives are recorded and preserved by hundreds of entities, often commercial in nature. The reason for this shift is simple: innovative companies have figured out how to deliver great Web-based services (services that also happen to make money) by divining clickstream patterns. Like most material culture, the clickstream is becoming an asset, certainly to the individual, but in particular to the Internet industry.

Some mine this asset by calculating patterns in the clickstream—Google's PageRank, for example[6]—and others take more direct approaches, such as the algorithms behind Amazon's recommendation system. Most visibly, all search engines mine clickstream data to present advertisements that attempt to match your stated intent.

From a consumer's point of view, there are also very simple and compelling reasons for this shift: services like search, recommendation networks, and e-mail make our lives easier, faster, and more convenient. We're willing to trade some of our privacy—so far, anyway—for convenience, service, and power.

"Search as a problem is about five percent solved," notes Udi Manber, the CEO of Amazon's A9.com search engine. Five percent—and yet the search business has already blossomed into a multibillion-dollar industry. Search drives clickstreams, and clickstreams drive profits. To profit in the Internet space, corporations need access to clickstreams. And this, more than any other reason, is why clickstreams are becoming eternal.

As we root around in the global information space, search has become our spade, the point of our inquiry and discovery. The empty box and blinking cursor presage your next digital artifact, the virgin blue link over which your mouse hovers awaits transformation into yet another imprint onto this era's eternal index.

Implications

What do Japanese teenagers think is cool this week? What pop star is selling, and who is falling off the charts? Which politician is popular in Iowa, New Hampshire, or California, and why? Where do suburban moms get their answers about cancer? Who visits terrorist-related or pornography sites, and how do visitors find them? What type of insurance do Latino men buy, and why? How do university students in China get their news? Nearly any question one might frame can be answered in one way or another by mining the implacable Database of Intentions that is building second by second across the Internet.

So what does the emergence of such an artifact augur? What effect might it have on the multibillion-dollar marketing and media industries? Why have the governments of China, Germany, and France threatened to ban search engines like Yahoo or Google, and why might our own national security hinge on plumbing the depths of their databases? What, in the end, might search tell us about ourselves and the global culture we are creating together online?

The answers to these questions are not simple, but I hope to at least address them as I tell the story of search in the pages that follow. Search straddles an increasingly complicated territory of marketing, media, technology, pop culture, international law, and civil liberties. It is fraught not only with staggering technological obstacles—imagine the data created by billions of queries each week—but with nearly paralyzing social responsibility. If Google and companies like it know what the world wants, powerful organizations become quite interested in them, and vulnerable individuals see them as a threat. Etched into the silicon of Google's more than 150,000 servers, more likely than not, are the agonized clickstreams of a gay man with AIDS, the silent intentions of a would-be bomb maker, the digital bread crumbs of a serial killer. Through companies like Google and the results they serve, an individual's digital identity is immortalized and can be retrieved upon demand. For now, Google cofounder Sergey Brin has assured me, such demands are neither made nor met. But in the face of such power, how long can that stand?

Eventually, such demand will surface, if, in fact, it has not done so already. The power of such a tool is staggering, and the threat of its being turned toward ill-considered ends quite real. In the aftermath of September 11, the Bush administration swiftly introduced legislation that redefined domestic surveillance powers. Swept up in the moment, Congress passed the USA PATRIOT Act[7] without debate. Under the act, the U.S. government may now compel companies like Google to deliver information to government agents on demand, and in secret.

The implications are far reaching, says Stewart Baker, former counsel for the National Security Agency (NSA). Under the PATRIOT Act, he told the *New York Times*, the government can demand information on "everyone you send e-mail to, when you sent it, who replied to you, how long the messages were, whether they had attachments, as well as where you went online." With entire divisions of the FBI, NSA, and Department of Defense now committed to Internet-based surveillance, databases as rich as AOL, Google, or Yahoo will not be overlooked. And given the fact that these companies are legally obligated to remain silent about what information they might give to the government, they are inherently conflicted between the government and their millions of trusting customers. As a Google executive noted to me when I brought this up: "We're one bad story away from being seen as Big Brother."

This reality raises interesting questions about privacy, security, and our relationship to government and corporations. When our data is on our desktop, we assume that it is ours. It's *my* address book that lives in Entourage, *my* e-mail attachments, and *my* hard drive inside my PowerBook. When I am looking for a file or a particular e-mail message on my local files (when I am searching my local disk), I presume that my mouse-and-click actions—those of searching, finding, and manipulating data—are not being watched, recorded, or analyzed by a third party for any reason, be it benign or malicious. (In many workplaces, this is certainly no longer the case, but we'll set that aside for now.)

But when the locus of computing moves to the Web, as it clearly has for second-generation applications like social networking, search, and e-commerce, the law is far fuzzier. What of the data that is stored and created through interactions with those applications?

Who owns that data? What rights to it do we have? The truth is, at this point, we just don't know.

As we move our data to the servers at Amazon.com, Hotmail.com, Yahoo.com, and Gmail.com, we are making an implicit bargain, one that the public at large is either entirely content with, or, more likely, one that most have not taken much to heart.[8]

That bargain is this: we trust you to not do evil things with our information. We trust that you will keep it secure, free from unlawful government or private search and seizure, and under our control at all times. We understand that you might use our data in aggregate to provide us better and more useful services, but we trust that you will not identify individuals personally through our data, nor use our personal data in a manner that would violate our own sense of privacy and freedom.

That's a pretty large helping of trust we're asking companies to ladle onto their corporate plate. And I'm not sure either we or they are entirely sure what to do with the implications of such a transfer. Just thinking about these implications makes a reasonable person's head hurt.

But imagine the disorientation you might feel if search becomes self-aware—capable of watching you as you interact with it.

Search as Artificial Intelligence?

"I would like to see the search engines become like the computers in *Star Trek,*" Google employee number one, Craig Silverstein, quips. "You talk to them and they understand what you're asking."

Silverstein, a soft-spoken paragon of Google's geek culture, is hardly kidding. The idea that search will one day morph into a humanlike form pervades nearly all discussion of the application's future. Asked at a conference how he'd best describe his search service, Ask Jeeves executive Paul Gardi replied: "[The android character] Data from *Star Trek*. We know everything you might need."

But how might we get there? For search to cross into intelligence, it must understand a request—the way you, as a reader, understand this sentence (one hopes). "My problem is not finding something," says Danny Hillis, a MacArthur Foundation genius and computer scientist who now runs a consulting business. "My problem is understanding something." That, he continues, can happen only if search engines understand what a person is really looking for, and then guide her toward understanding that thing, much as experts do when mentoring a student. "Search," he continues, "is an obvious place for intelligence to happen, and it is starting to happen."

So Hillis argues that the future of search will be more about understanding, rather than simply finding. But can a machine ever understand what you are looking for? Answering that question raises what is perhaps computing's holiest of grails: passing the Turing test.

The Turing test, explained by British mathematician Alan Turing in a seminal 1950 article, lays out a model to prove whether or not a machine can be considered intelligent. While the test and its prescripts are subject to intense academic debate, the general idea is this: an interrogator is blindly connected to two entities, one a machine, the other a person. The questioner has no idea which is which. His task is to determine, through questioning both, which is human and which is machine. If a machine manages to fool the questioner into believing it is human, it has passed the Turing test and can be considered intelligent.

Turing predicted that by the year 2000, computers would be smart enough to have a serious go at passing the Turing test. He was right about the serious go part, but so far, the prize has eluded the best and brightest in the field. In 1990, a wealthy oddball, Hugh Loebner, offered $100,000 to the first computer to pass the test. Every year, AI companies line up to win the honor. Every year, the money remains uncollected.

That may well be because, as with so many things, people are framing the problem in the wrong way. So far, contestants have focused on building singular robots that have millions of potential answer sequences coded in, so that for any particular question a plausible answer might be given.[9] Perhaps the most famous of these efforts is Cyc (pronounced "psych"), the life's work of AI pioneer Doug Lenat. Cyc attempts to conquer AI's brittleness problem by coding in hundreds of thousands of commonsense rules—mountains go up, then down, valleys are between hills or mountains, and so forth—and then building a robust model based on those simple rules. Not surprisingly, a Cyc alumnus, Srinija Srinivasan, was one of Yahoo's first employees, and has run Yahoo's directory-based search product from nearly day one.

But brute force by one organization has failed so far, and most likely will fail in the future. No, search will more likely become intelligent via the clever application of algorithms that harness and leverage the intelligence already extant on the Web—the millions and millions of daily transactions, utterances, behaviors, and links that form the Web's foundation—the Database of Intentions. After all, that's how Google got its start, and if any company can claim to have created an intelligent search engine, it's Google.

"The goal of Google and other search companies is to provide people with information and make it useful to them," Silverstein tells me. "The open question is whether human-level understanding is necessary to fulfill that goal. I would argue that it is."

What does the world want? Build a company that answers this question in all its shades of meaning, and you've unlocked the most intractable riddle of marketing, of business, and arguably of human culture itself. And over the past few years, Google seems to have built just that company.

Chapter 2
Who, What, Where, Why, When, and How (Much)

Judge of a man by his questions, rather than by his answers.
—Voltaire

Before we take a long journey around the contours and implications of search, it makes sense to get our bearings. Back when I was a cub reporter, I was taught to answer five questions about any topic before writing about it: who, what, where, why, and when. If you crammed answers to all those questions into your lead paragraph, then you'd essentially done your job.

But to those five questions I quickly learned to add a sixth—how?—and a corollary: who's making the money, and how much? We'll get to the money question last, but first, let's address the how.

How

So how does a search engine work? There's a very, very long answer to this question, but I'll stick to a shorter one. In essence, a search engine connects words you enter (queries) to a database it has created of Web pages (an index). It then produces a list of URLs (and summaries of content) it believes are most relevant for your query. While there are experimental approaches to search that are not driven by this paradigm, for the most part, every major search engine is driven by this text-based approach.

A search engine consists of three major pieces—the crawl, the index, and the runtime system or query processor, which is the interface and related software that connects a user's queries to the index. The runtime system also manages the all-important questions of relevance and ranking. All three pieces are integral to the quality and speed of the engine, and there are literally hundreds of factors in each that affect the overall search experience delivered. But the basics are pretty much the same for all the engines. As Tim Bray, a search pioneer now at Sun Microsystems, puts it in his excellent series "On Search," "The fact of the matter is that there really hasn't been much progress in the basic science of how to search since the seventies."

The search all starts with you: your query, your intent—the desire to get an answer, find a site, or learn something new. Intent drives search—a maxim I'll be repeating time and again throughout this book. We'll get into the query a bit more in the "What" section below, but on average we enter one or two short words into a query box each time we search, and we click on an average of two or so results among the millions an engine often lists. In addition, the average Web searcher conducts about one search a day. Of course, that's an average. A small percentage of hopelessly connected surfers conduct hundreds of searches a day, and many more do no more than one or two a month. (All these figures, as one might expect, are growing over time.)

The process of how we get our results starts with the crawler. The crawler is a specialized software program that hops from link to link on the World Wide Web, scarfing up the pages it finds and sending them back to be indexed. It's seductive to think of crawlers as tiny little robots wandering the vast halls of cyberspace, but the truth is a bit more mundane. Crawlers are in fact homebodies, sitting on their own servers and sending out vast numbers of requests to pages on the Internet, much as your browser does.

Those requests bring back Web pages, which the crawler then hands off to the indexer. It also takes note of any links it has found on the page, and queues those links in its request file—sending out yet more requests to the newly found links, which find more links . . . and so on, ad infinitum. Though the science behind crawlers is complex, what they do is pretty simple: they go off on a endless binge of dialing for URLs, and they report back what they've found. Crawlers have long been the least visible of the search engine's components, but they are arguably the most important. The more sites they crawl, and the more frequently they crawl them, the more complete the index is. When the index is more complete, the search results pages (SERPs) that are returned for a particular query have a greater chance of being relevant.

Early versions of crawlers discovered and indexed only the titles of Web pages, but today's more advanced versions index the contents of the entire Web page, as well as many different file types such as Adobe Acrobat (PDF), Microsoft Office documents, audio and video, and even site-specific metadata—structured information provided by site owners about the pages or information being crawled.

The crawler sends its data back to a massive database called the index. The index breaks into several pieces, depending on whether the data has been processed and made ready for consumption by searchers like you and me. Raw indexes are rather like lists organized by domain: for any given site, the index will list all the pages on that site, as well as all pertinent information about those pages: the words on the page, the links, the anchor text (text around and within a link), and so on. The information is organized in such a way that if you know the URL you can find the words that are related to that URL.

Why is this important? Because the next step in creating a smart index is to invert the database—in essence, to make a list of words that are then associated with URLs. So when you type "outer Mongolia" into a search box, the engine immediately can retrieve a list of all the URLs that include those words.

The first engines on the Web essentially executed to this point, and not much further. But since the late 1990s, the index has become a significant area of innovation for all search companies—where much of a search engine's secret sauce is applied.

Think of the index as a huge database of important information about Web sites. Innovative companies like Google have made their reputation by studying that database—noting statistical patterns and algorithmic potentials, divining new ways to leverage it toward the ultimate goal of providing you with more relevant results for your queries.

The process of grokking the index is referred to as analysis. Google's PageRank algorithm is an example of analysis: it looks at the links on a page, the anchor text around those links, and the popularity of the pages that link to another page and factors them together to determine the ultimate relevance of a particular page to your query. (While PageRank is often understood to be an "all-knowing" algorithm, Google, in fact, looks at more than one hundred factors to determine a site's relevance to your keywords.)

Through the process of analysis, indexes are populated with tags, another kind of metadata—data about data. Pages might be tagged as written in a certain language, for example, or as belonging to a certain group such as porn, spam, or rarely updated. This metadata is critical to an engine's ability to offer you the most relevant results.

Once the crawl data is analyzed, indexed, and tagged, it's dumped into what's called a runtime index—a database ready to serve results to users. The runtime index forms something of a bridge between the back end of an engine (its crawl and index) and the front end (its query server and user interface).

The query server is software that transports a user's search query from the user interface—the home page of search.yahoo.com, for example—to the runtime index, then shuttles SERPs back to the interface. While much of an engine's intelligence is built into analysis, the query server can hold quite a bit as well. If you've spent any time banging on different types of search engines, you can see some of that front-end intelligence at sites like Ask.com, which clusters its results around various flavors of possibly relevant topics. Search on Ask.com for "jaguar," and you'll be given a list of related searches that attempt to narrow your search. Did you mean "animal jaguar," or "car and jaguar"? Many engines use interface tricks like this one to aid searchers in their quest for the right result.

At the end of day, the holy grail of all search engines is to decipher your true intent—what you are looking for, and in what context. And while search engines are increasingly getting better at this task, they are nowhere near solving this problem. An example of progress in this area is in the identification of what are called atomic phrases. When you type in a one-word query for "York," for example, do you want results for "New York"? Most likely the answer is no. In the past two years, most engines have evolved to tell the difference by parsing a list of atomic phrases—phrases that have their own sets of results at the smallest levels.

As search users, we are extraordinarily good at incoherence, making the task of procuring useful search results a Herculean task. You and I know what we mean when we type "Abraham Lincoln biography" into a search box, for example. You aren't necessarily looking for every page that has those words on it, but rather pages that conceptually can be understood to contain biographies of the famous president. But how might a search engine understand such a concept? One way is by the use of cue words that tip the engine off to the context of a particular search. In this case biography is a concept, not just a word that might be found on a page. A good query engine will link this cue word to clusters of results that have a chance of fulfilling the concept of biography—pages that have been tagged as biographical. Adding that new metadata often dramatically improves results. (Other examples of cue words or phrases include "movie reviews," "stock quotes," and "weather reports.")

In a similar vein, engines must deal with local variances and the problem of a lack of a controlled vocabulary. Nearly all programming languages employ a very strict grammar in order to communicate between humans and machines. If one comma is out of place or one word misspelled, the program will fail. Search can't afford such strictures, and search engines are still working on the problem of how to best match searches for "soda" with results for "pop," "tennis shoes" with "sneakers," or "feline" with "cat."

Search engines also do better by doing less: most engines have a list of stop words that are ignored—common words with little semantic value such as "to," "the," "be," "and," and "or." Tossing out these words saves the indexes valuable processing cycles, but makes a search for the phrase "to be or not to be" something of a wild goose chase.[1]

Search companies obsess about these and other patterns in the clickstream of search. They watch what you search for, what results you choose to click on, and even where you go after that so as to determine better algorithms to apply to results pages. "You can learn a lot by watching the statistical patterns of search usage, and leveraging that in algorithms," notes Gary Flake, the former head of Yahoo's research labs, who now works for Microsoft. "We use a very large corporea [body of data] to identify sets of tactical and grammatical properties of language." The result: search has the potential to get better and better, the more people use it. A good example is the spell checker found at Google and other major search engines—its suggestions are culled from watching vast numbers of misspellings and correlating them to the properly spelled word.

To summarize, there are three critical pieces of search, and all three must scale to the size and continued growth of the Web: they must crawl, they must index, and they must serve results. This is no small task: by most accounts, Google alone has more than 175,000 computers dedicated to the job. That's more than existed on Earth in the early 1970s!

Finally, in addressing the "how" of search, it's important to take a quick detour into the specific methods we as searchers deploy. The short of it is this: we are incredibly lazy. We type in a few words at most, then expect the engine to bring back the perfect results. More than 95 percent of us never use the advanced search features most engines include, and most search experts agree that the chances of ever getting that number lower are slim to none. We want results now, and we want that engine to provide them without forcing us into learning an unwieldy new programming language (although unquestionably, search is shaping our cultural grammar in ways we have yet to understand).

But a quick study of common advanced search tricks will yield significantly better results. Most engines offer the ability to narrow a search by phrase, domain, file type, location, language, and number of results. You can include or exclude keywords, set specific time frames for results, and, with many engines, even search for pages that are similar to those you find useful. It's beyond the scope of this book to teach advanced search techniques, and honestly, I'm as lazy as most when it comes to using them. But if you're looking to learn more, there's plenty of help out there.[2]

Who

Moving back to the original set of questions, let's tackle the "who." Who searches the Web? The simple answer is nearly everyone, but that certainly isn't a satisfying answer. We can learn quite a bit from the data collected so far on search habits. In the summer of 2004, the Pew Internet & American Life Project released a research paper on American usage of the Internet (we'll tackle international usage in a minute). It concluded that of all Americans who use the Internet, about 85 percent use search engines, or more than 107 million people in the United States alone. More than two-thirds of those are active users of search—employing one search engine or another more than twice a week and averaging more than thirty searches a month.

Pew estimates that on any given day in the United States, 38 million people are using a search engine. All those searches add up to nearly 4 billion queries each month. And those are just queries on the Internet's most popular search engines—they don't include the search boxes of Amazon.com, eBay, or the thousands of smaller search-driven businesses and information sites. Only e-mail is a more popular online tool, the project concluded. And according to research from investment bank Piper Jaffray, search usage continues to grow—on average by nearly 20 percent per year—with the majority of that usage growth driven by new search users. The number of searches per user is also increasing, by about 25 percent per year.

So who are these people, the folks using search engines? Are they any different from the average American? Turns out the answer is yes. Pew has found a technology elite that drives usage of the Internet. Thirty-one percent of the U.S. population, Pew claims, are members of this elite. Pew also found that the younger you are or the higher your educational attainment is, the more you search. An interesting corollary: as we search more, we are also becoming more connected, more digital, and more dependent on information services: the household spending for media and information services in the United States rose at an annual rate of 32 percent throughout the 1990s, from $365 a year to $640.

What

Now that we've established who is searching and how the process works, what are people searching for? Therein lies the beauty and the potential of search: it's driven by the unimaginable complexity inherent in human language—nearly infinite combinations of dialects, words, and numbers. Piper Jaffray estimates that the world conducted about 550 million searches each day in 2003, a figure it expects to grow at about 10 to 20 percent a year. NetRatings, a U.S.-based research company, estimates that U.S. searches are growing even faster—by 30 percent a year. That means from the time these words are written to the time this book sees print, total queries in the United States will have risen from 4 billion a month to well over 5 billion—an astonishing rate of growth.

As I mentioned in the "How" section above, the query is the

lodestone of search, the runes we toss in our ongoing pursuit of the perfect result. According to a June 2004 Majestic Research report, as searchers we are a rather terse lot. Nearly 50 percent of all searches use two or three words, and 20 percent use just one. Just 5 percent of all searches use more than six words. Overall, though, the trend is toward adding more girth to our queries as we navigate this odd new grammar of the keyword.[3]

But focusing on the number of words in a search query misses the point: it's not the complexity of the search that matters; it's the complexity of our language.

Thorstein Veblen, the early-twentieth-century thinker who coined the term "conspicuous consumption," once quipped "The outcome of any serious research can only be to make two questions grow where only one grew before." As anyone who has spent an afternoon in a fruitless search can attest, coming up with the right words to find what you're looking for can be a frustrating task. You know there's an answer out there, but you just can't seem to come up with the right combination of words to find it. In fact, Pew research shows that the average number of searches per visit to an engine is nearly five. Clearly we are not getting what we want the first time or we're coming up with new questions driven by the results our initial questions return.

Arguably, there is no greater act of creativity than the formation of a good question, and every day the wired world asks hundreds of millions of questions via search. While it's tempting to conclude that we all ask pretty much the same questions, in fact the truth lies somewhere in between. We do ask a lot of the same questions, but we ask far more that are unique, and therein lies the power of search.

If you were to plot a list of a thousand random queries along a horizontal line, and then plot their frequency up a vertical one, you'd come up with a graph that looked a lot like the one on p. 28.

In other words, there are a few queries that have very high frequency, but quickly the graph flattens out into a massive tail, a tail that is extraordinarily long. And the power of search lies in that tail: no matter what the word is, somewhere on the Web there's most

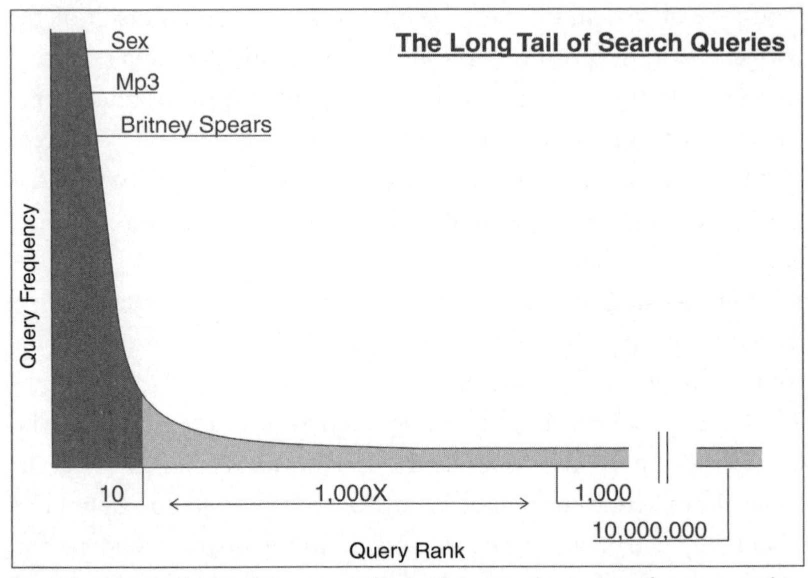

Average query frequency for query ranks 1–10 trumps the average for query ranks 11–110 a thousandfold. Source: Joe Kraus.

likely a result that contains it. According to Piper Jaffray research, each day more than 50 million unique keyword combinations are entered into search engines in the United States. And Google puts the figure much higher: it claims that nearly 50 percent of the searches coming in on any given day—more than 100 million—are unique. (In fact, in the early days of Google, a popular sport among search watchers was to find a query that had exactly one result. This game even has a name—GoogleWhacking.) This copious diversity drives not only the complexity of search itself, but also the robustness of the advertising model that supports it: there are literally millions of keywords to purchase that just might have economic value to someone, at some time.

But as with all things one can generalize search queries into large categories. According to Piper Jaffray, while 20 percent of searches are for entertainment information and 15 percent are commercial in nature, the majority, 65 percent, are informational. According to the Kelsey Group, as much as 25 percent of all searches are local, and most of those are commercial in nature (looking for a dentist, a restaurant, a plumber).[4]

And according to a 2004 Harris poll, nearly 40 percent of us have done a vanity search—typed our own name into a search engine to see if we exist in the doppelgänger of the search index. I'd be willing to wager that this number will head north of 90 percent in the coming years, as search becomes as individually definitional as finding oneself in the white pages was during the rise of the telephone. Besides ourselves, nearly 20 percent of us have looked for former flames and 36 percent for old friends, and 29 percent have researched a family matter.

An older but still relevant academic paper gives us a few clues as to what we really are looking for. *A Taxonomy of Web Search* by Andrei Broder, written largely while the author was CTO of AltaVista in 2001, was based on query data from that early innovator in search. Broder sets out to dispel the notion that most searches are informational in nature. He instead maintains that many are transactional or navigational.

A few fun facts from Broder's analysis of response and related log data:

- Nearly 15 percent of searchers wish for "a good collection of links on a subject" as opposed to "a good document."
- Queries that were sexual in nature make up 12 percent of the log data.
- Nearly 25 percent of searchers were looking for "a specific Web site that I already had in mind."
- An estimated 36 percent of searchers were looking for transactional information—what Broder calls "the intent to perform some Web-mediated activity."

That Web-mediated activity translates into commercial searches, though the difference between commercial searches and informational ones is not as clear as might be expected. In fact, Piper Jaffray's data suggests that the true percentage of commercial searches on the Net is more than 35 percent. On the Internet, it can be argued, all intent is commercial in one way or another, for your very attention is valuable to someone, even if you're simply researching your grandmother's genealogy, or reading up on a rare species of dolphin. Chances are you'll see plenty of advertisements along the way, and those links are the gold from which search companies spin their fabled profits.

Where, Why

So far we've reviewed how search works, who is searching, and what people are searching for. But where are they going, and why are they going there in the first place? In the aggregate, most searchers stick close to home: 85 percent use one of the big four portals—Microsoft, Yahoo, Google, or AOL.[5] And they tend to stick with these engines once they've started: market share among the giants has fluctuated in the past years, but even with major moves by both Microsoft and Yahoo to improve their search results, Google remains the market leader.

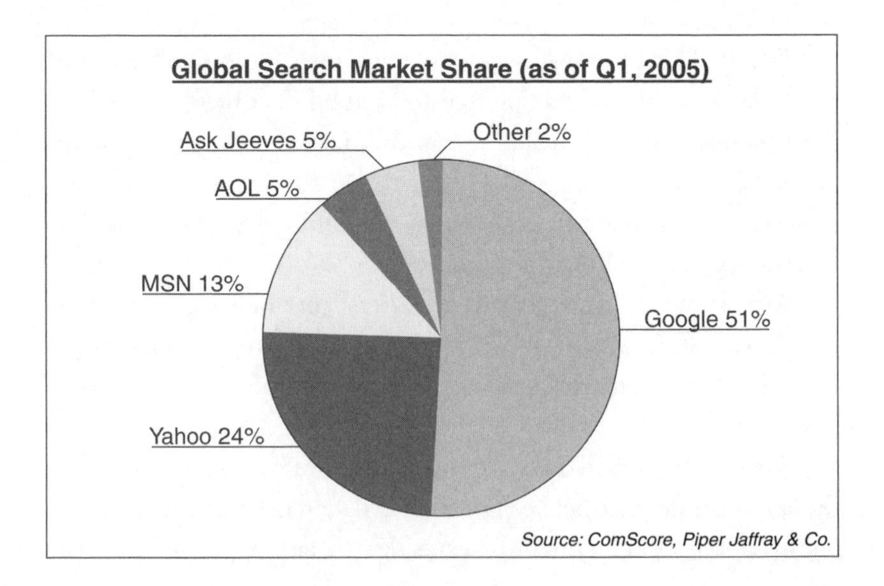

While Internet penetration in the United States is more than ten times the average for the rest of the world, far more searches are done internationally than in the United States—by a factor of more than five to one. For this reason, one can argue that if you wish to understand the future of search, you'd better learn to speak another language or two.

As to the question of why we search, aside from securing our immortality, the answer is more complicated that it might seem. Sure, we search to find information on all manner of things, or to locate something to buy, or to simply find the shortest route to a site we already know exists (the practice of typing in a word you know so as to yield a site you wish to visit, also called a navigational query). In short, we search to find.

"The 'why' of user search behavior is actually essential to satisfying the user's information need," write Yahoo researchers Daniel E. Rose and Danny Levinson in a paper entitled "Understanding User Goals in Web Search." "After all, users don't sit down at their computer and say to themselves, 'I think I'll do some searches.' Searching is merely a means to an end—a way to satisfy an underlying goal that the user is trying to achieve. (By 'underlying goal,' we mean how the user might answer the question 'why are you performing that search?') That goal may be choosing a suitable wedding present for a friend, learning which local colleges offer adult education courses in pottery, [or] seeing if a favorite author's new book has been released."

In other words, we are searching for more than answers. Not only are we searching for that which we know; we increasingly are searching to find that which we do not know, a state similar to searching in the initial stages of the Internet, when no one knew what was out there. As Jerry Yang of Yahoo tells me, back when he started the service as a directory, no one knew what was out there, and a directory listing cool new sites was a revelation. But our need to comprehend what was out there receded as we began to know our way around—now we assume that everything is connected. That vastness is causing another kind of Web blindness: a sense that we know there's stuff we might want to find, but have no idea how to find it. So we search in the hope it will somehow find us.

Jeff Bezos, CEO of Amazon, calls this kind of searching discovery: the process of casting about to encounter that which we hope might find us. (Bezos has made quite a business of discovery-based search. Amazon's "people who bought your product also bought . . ." recommendation system is one of the company's most profitable secret weapons.) Indeed, many in the industry make what I think is an important distinction when it comes to search: there is search to recover that which we know exists, and then there is search to discover what we intuit exists, but have yet to find. In this book, when I refer to search in its most general terms, I intend the word to include both recovery and discovery.

So why do we search? To recover that which we know exists on the Web, and to discover that which we assume must be there, be it a pottery class or a long-lost friend.

When

The rather mundane question of when can be boiled down to one straightforward fact: we search from both home and work, with our searches pretty much evenly broken up between them. Search traffic tends to increase in the morning and peaks again in the evening, as we all fire up our home computers and look for movie tickets, homework help, or a local plumber to fix the dripping sink.

I'll take the "when" question historically and use it as an excuse to provide some context as to how we got to the present day in search. Humankind has searched for archived information ever since the dawn of symbolic language; the index and the archive are as ancient as the clay tablet. The technology of classification and information retrieval (IR), as the academic domain is known, did not really take flight until the rise of the printing press and the resultant explosion of widely available printed matter.

In the late nineteenth century Melvil Dewey, widely credited as the father of the modern library, introduced a universal classification system based in large part on a directory-like structure that identified books by their subject using a numeric code. The Dewey decimal system has been updated numerous times over the years and is still widely used, but its subject-based focus would be unable to scale to the enormousness of the World Wide Web.

The "when" of Internet search can be traced to the rise of the digital computer in the 1940s and 1950s. As the computer began to take over back-office functions like inventory, payroll processing, financial calculations, and academic research, institutions started to collect large amounts of data, data that, because of the peculiar nature of digital computing, was searchable. This breakthrough led to a revolution in the field of information retrieval. How might one classify information in its most atomic form—the word—as opposed to a book or pamphlet?

Enter Gerard Salton, a Harvard- and Cornell-based mathematician often called the father of digital search. Salton was fascinated by the problem of digital information retrieval, and in the late 1960s developed SMART—Salton's Magical Automatic Retriever of Text—or what might be considered the first digital search engine. Salton introduced many of the seminal concepts commonly used in search today, including concept identification based on statistical weighting, and relevance algorithms based on feedback from queries. Salton's work sparked a renaissance in the IR field and inspired an annual conference on digital information retrieval known as the Text Retrieval Conference (TREC).

From the early 1980s to the mid-1990s, TREC reflected the state of the art in text search. Academics and researchers gathered to test each other's mettle in finding the most relevant results from a standardized body of news articles. But TREC largely ignored the early Web—it was simply too unruly and unpredictable. As Google founders Larry Page and Sergey Brin wrote in the paper that announced Google to the academic community in 1997: "The primary benchmark for information retrieval, the Text Retrieval Conference (TREC 96), uses a fairly small, well-controlled collection for [its] benchmarks. The 'Very Large Corpus' benchmark is only 20GB compared to the 147GB from our crawl of 24 million Web pages. Things that work well on TREC often do not produce good results on the Web. . . . Another big difference between the Web and traditional well-controlled collections is that there is virtually no control over what people can put on the Web. Couple this flexibility to publish anything with the enormous influence of search engines to route traffic and companies which deliberately manipulat[e] search engines for profit become a serious problem. This [is a] problem that has not been addressed in traditional closed information retrieval systems."[5]

Page and Brin go on to describe their solution to text retrieval on the Internet, and the rest, as they say, is history. (Well, almost. For an overview of the world of Internet search pre-Google, head to Chapter 3).

The Money Shot

All those searches, and all those searchers, have translated into a major business opportunity, in fact, the fastest growing business in the history of media. From its inception as a business in the late 1990s to 2004, paid search as an industry grew from a base in the low millions to $4 billion in revenue, and it is estimated to hit $23 billion by 2010, according to Piper Jaffray. With numbers like that, it's no wonder Google's IPO rocketed to $200 a share in its first six months of trading.

Why the extraordinary growth? In short, paid search works. Lining up brief, text-based advertisements against the queries of those hundreds of millions of searchers results in extremely efficient marketing leads, and marketing leads are the crack cocaine of business. Marketing leads, for those of you who prefer your English in nonbusiness terms, are inquiries from potential customers. All those CDs in your mailbox from AOL? All the junk mail from Publishers'

Clearinghouse? The unwanted phone calls from your bank during dinner? Each one of them is an attempt by a business to garner a marketing lead, the most sought-after source of new business in the Western economy. So why is search so hot? Take a look at this chart from Piper Jaffray:

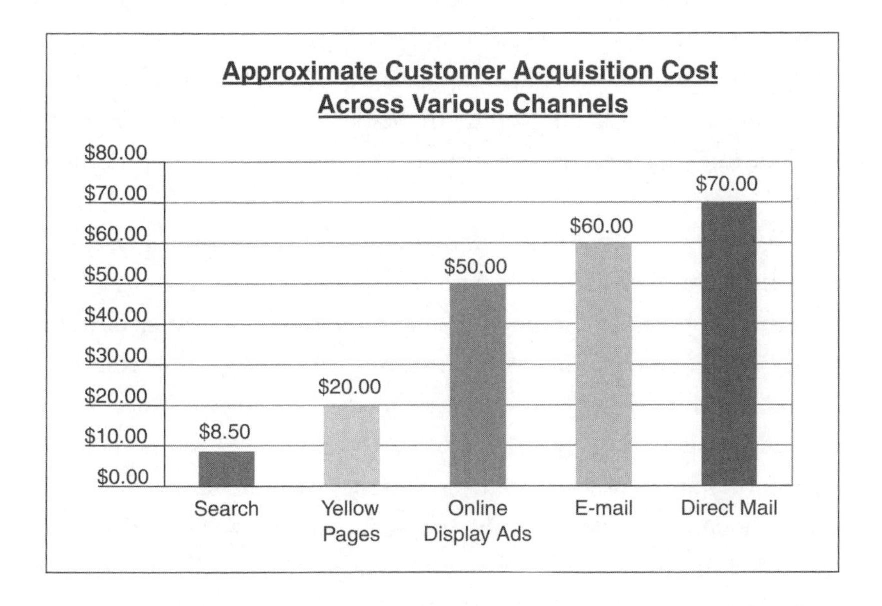

That just about says it all. Search, a marketing method that didn't exist a decade ago, provides the most efficient and inexpensive way for businesses to find leads. In the past five years, the number of unique advertisers who have implemented search marketing programs has grown from the thousands to the hundreds of thousands. Google alone boasts more than 225,000 unique advertiser relationships. Try that with network television![6]

About 40 to 50 percent of all search queries now return paid ads alongside the results, according to Majestic Research, and that number will only increase over time as companies optimize their sites to convert searchers to paid clicks. Once those sponsored links appear, 13 to 14 percent result in conversion to a paid click, according to Majestic (these figures are an average for Google and Yahoo only).

That's not much, one might argue, until one does the math. The average price per paid click was hovering at about 50 cents in early 2005. Between Google and Yahoo, there are more than 2 billion searches each month. Back of the envelope: 2 billion times 14 percent—that's about 280 million paid clicks. Multiply that by an average of 50 cents and you have about $140 million in revenue each month to split between the two. And that's just on their home pages. Both Yahoo and Google have extensive networks serving other sites, providing a similar if not slightly higher level of traffic and revenue. Bottom line: all those clicks add up to billion-dollar revenue lines for both companies.

Why do so many folks click on paid ads? Not surprisingly, there are a huge number of people who use the Web to research and buy things. According to a report from the Dieringer Research Group, nearly 100 million people made purchases after doing online research in 2003, and nearly 115 million searched for product information.

And while Google and Yahoo are the dominant forces in paid search, they are in no way alone, nor do they own the innovation that such a booming market spawns. While the first phase of paid search depended almost exclusively on the concept of matching text ads with the intent of a search query, second- and third-generation search advertising models are emerging, and any number of them might again fuel a major upswing in spending. Currently, most of the major players are eyeing the local search market, which at present is served not by the engines but by a decidedly offline medium: the yellow pages. At the time of this writing, the local search business is measured in the hundreds of millions, but the yellow pages is a $14 billion business in the United States ripe for the picking. Ask, Yahoo, Google, Citysearch, and many smaller players have all launched local search products, and the yellow pages companies have responded with online services of their own. Their bet: that soon the local dentist, restaurant, or dry cleaner might best spend his $500 on a search engine, instead of a listing in the local yellow pages.

Besides looking for new market segments like local, search companies and new start-ups are focusing on several innovative approaches to monetizing your clickstream. Behavioral targeting, for example, seeks to track your search and browsing history and display advertisements that might be contextually relevant based on your online behaviors. Similarly, search personalization attempts to determine who you are, by either demographic data you provide (as is the case when you register at Yahoo) and/or by your clickstream history. This way, an engine can provide more relevant results, as well as more highly targeted ads. If, for example, you seem to be looking for "Lincoln" quite a bit lately, and tend to click not on results related to the president, but rather on the automobile, second-generation engines will display ads for Lincoln cars (or, as is often the case, ads for competitors to Lincoln).

As the search economy deepens and proliferates, there will be countless innovations built upon the basic breakthrough of the paid search model. But before we head into the economic implications of Web search, or the story of Google, its brightest star, it's wise to spend a little time considering a bit of history. For while it seems that the words "Google" and "search" are now nearly one and the same, the truth is that search has been around for decades, in one form or another. Google is currently our culture's grandest declaration of the power of search—but it is by no means the first.

Chapter 3
Search Before Google

AltaVista wasn't first, but they were first to do it in a way that could be considered a significant improvement over state of the art.

—Dr. Gary Flake, distinguished engineer, Microsoft Corp.

Early Search

By most accounts, the honor of being the first Internet search engine goes to Archie, a pre-Web search application created in 1990 by a McGill University student named Alan Emtage. By 1990, academics and technologists were regularly using the Internet to store papers, technical specs, and other kinds of documents on machines that were publicly accessible. Unless you had the exact machine address and file name, however, it was nearly impossible to find those archives. Archie scoured Internet-based archives (hence the name "Archie") and built an index of each file it found.

Based on the Internet's file transfer protocol (FTP) standard, Archie's architecture was similar to most modern Web search engines—it crawled sources, built an index, and had a search interface. But the pre-Web era was not a very user-friendly time. Only true techies and academics used Archie, though among that crowd it was quite popular. Typical users would query the engine by connecting directly to an Archie server via a command-line interface. They would query Archie via keywords thought to be in a matching file's title, then receive a list of places where a particular matching file could be found. They then connected to that machine, and rummaged around till they found what they were looking for. Not particularly robust, but much better than nothing.

The name "Archie" had a quirky appeal that seemed to fit the young Internet. In 1993, students at the University of Nevada created Veronica, a play on the comic book couple. Veronica worked much as Archie did but substituted Gopher, another popular and more fully featured Internet file-sharing standard, for FTP. Veronica moved search a bit closer to what we now expect—the Gopher standard allowed searchers to connect directly to the document queried, as opposed to just the machine on which the document resided. Not a huge step, but progress.

Both Archie and Veronica lacked semantic abilities—they didn't index the full text of the document, just the document's title. That meant a searcher had to know—or infer—the title of the document he or she was looking for. If you were looking for a "to-do list" and its title was "Today's Tasks" you'd be out of luck, even if the document's first words were, in fact, "to-do list." With the rise of the Web, Archie and Veronica soon fell out of favor.

As the Web took off, so did the basic problem of search. When the Internet was the domain of academics and technologists, finding things was a limited problem. But from 1993 to 1996, the Web grew from 130 sites to more than 600,000. Watching all this growth was Matthew Gray, a researcher at the Massachusetts Institute of Technology and a pioneer of the earliest Web-based search engine, the WWW Wanderer.

The Wanderer solved a basic problem Gray had noted with the Web, namely, that it was growing faster than any human could track. "I wrote the Wanderer to systematically traverse the Web and collect sites," Gray later wrote. "As the Web grew rapidly, the focus quickly changed to charting its growth."

The Wanderer was a robot that automatically created an index of sites, and Gray hacked up a search interface that allowed users to search the index. Gray's Wanderer had another, unexpected effect: in the early days of the Web, bandwidth was at a premium, and many Webmasters felt the Wanderer ate up too many processing and bandwidth cycles as it indexed a site's contents. Gray later tweaked the crawler, setting it on a breadth algorithm to span many sites before drilling down—a more efficient process that's still used today. "It wasn't the greatest search engine that ever was, but it was the first search engine that ever was," Gray says.

The Wanderer was soon eclipsed by more powerful engines. One of the first was WebCrawler, developed by University of Washington researcher Brian Pinkerton. Pinkerton stumbled onto WebCrawler while working for Steve Jobs's company Next in 1994. (Jobs's Next machine and its NextStep software were, as were the products of so many pioneering companies, about five years ahead of the market. The technologies the company developed—built-in Ethernet, high-quality color—are now de rigueur in nearly every desktop PC). At the time, Pinkerton was juggling his academic work—molecular biotechnology and computer science—with his day job, where he was tasked with building a next generation Web browser with built-in search features for the NextStep operating system.

Pinkerton grew intrigued with search and the technology required to index the Web. It was an easy leap to make: a Web crawler retrieves URLs in much the same fashion as a Web browser. Pinkerton created a rudimentary crawler and started indexing Web sites.

Foreshadowing the importance of links and the eventual rise of Google's PageRank algorithm, Pinkerton then ran a test against his newly created database in March 1994. Which sites, he wondered, had the most references, or links, from other sites (in today's parlance, the most Googlejuice)? Number one on his list: the home page of the World Wide Web project at CERN, a major particle physics laboratory in Geneva, Switzerland.

WebCrawler was important to the evolution of search because it was the first to index the full text of the Web documents it found. Pinkerton put his extracurricular project online in April 1994. By November, it had recorded its one-millionth query—Pinkerton reports that it was "Nuclear Weapons Design and Research." In June 1995, AOL, which at that time had no Web-related assets, acquired WebCrawler for around $1 million, a portent of the search-related acquisition spree to come. "Nobody had an inkling of what the Internet would become," says Pinkerton.

WebCrawler opened up a new universe for Web surfers, particularly at AOL. Its full-text search and simple browser-based interface was an important step toward making the Web fit for mainstream consumption, beyond academics and tech geeks.

The First Truly Good Search Engine

When the Internet was young, when the Web comprised less than 10 million pages, when Yahoo was a funky set of links and "google" was just a common misspelling for a very large number, Louis Monier put the entire Web on a single computer.

There is a legend about the founding of AltaVista.com that goes something like this: Digital Equipment Corp. (DEC) had just come out with its superfast Alpha processor and was looking for some way to prove its might. Since massive databases lay at the heart of the corporate IT market, DEC needed a massive database to search. As the company was struggling and bereft of good news, it also needed a compelling PR angle to play up, something that might help it recapture its reputation as a technology innovator. Louis Monier, a researcher at DEC's Western Lab in Palo Alto, California, suggested building a search engine: it could load the entire Internet (the massive database) onto an Alpha computer, then build a program showcasing

Alpha's speed (that would be the search engine). Presto—AltaVista was born, a proof point to DEC's hardware dominance. But as with most founding stories (eBay's Pez dispenser mythology comes to mind[1]), the story is only half true.

"It was an after-the-fact rationalization," Louis Monier declares. Monier does not mince words as he recalls the early days of AltaVista. "DEC was in a death spiral," he tells me over coffee at a Palo Alto café. "They had screwed up any number of things." As his name implies, Monier is French; his inflection and delivery are pretty much central casting for Gallic contempt. "Nobody inside DEC understood what I was doing," he continues. "They were professionals of the missed opportunity. . . . They just thought it would make for a great demo for the hardware story."

In fact, plenty of folks inside DEC understood what Monier was up to, but unfortunately most of them were in the research division. And the story of AltaVista's birth will vary depending on whom you speak to. Brian Reid, who ran DEC's Network Systems Lab in the early 1990s, certainly remembers Monier's role in founding AltaVista. It was in Reid's staff meeting one morning that the idea for a search engine sprang up, he claims. Monier was there, and Monier took the idea and ran with it.

Monier may have gotten the credit, but "AltaVista was born in my conference room," Reid claims. "We were trying to figure out ways to use our extraordinary bandwidth. We had the new chip, a lot of smart people, and a company that was failing. We wanted to find a hook for the new machine, something that it could do better than HP or Sun." In Reid's telling, the legendary version of the story is pretty much on target.

When I ask Reid if Monier's version is correct, he pauses before admitting that in the end, no one really knows how the engine really came to be. "There is a lot of historical dispute about that," Reid admits. "There was a huge amount of backstabbing to take credit for the idea." At large companies like DEC, Reid explains, everyone wants credit for an idea that actually works, one that, in fact, makes the company look good. And for a brief moment, AltaVista was such an idea.[2]

As is true for much of the IT industry, nearly every well-known company in search can trace its roots to a university, the kind of institution that allows big ideas to flourish without the straitjacket of commercial demands. Google, Excite, and Yahoo emerged from Stanford; Inktomi came from University of California, Berkeley; and Lycos came from Carnegie Mellon.

Every so often a great innovation will spring not from a university, but from within a corporation. A few technology companies understand and nurture the ethos of academic research—open inquiry, freedom to fail, research without resource constraints, and open collaboration.

But not many companies can afford the luxury of pure research labs, and even fewer have the foresight and long-term vision to create them. But those that invest in pure research do so with a singular belief: that the innovations fostered by the research lab's fertile soil might someday provide the company a bridge to the future, safe passage across the treacherous crosscurrents of a hyperkinetic industry. Then, of course, there's the lottery play: theoretically, pure research allows for great leaps forward, leaps that may contain within them the spark of a hundred-billion-dollar opportunity. Not that that's the *stated* purpose of pure research, of course. But a company can dream.

Back in the late 1980s, DEC was among the few IT giants making a long-term investment in pure research. And for a moment in time, its premier laboratory, the Western Research Lab in Palo Alto, California, offered such a bridge to the future in the guise of a search application called AltaVista.

Xerox Corp may get all the blame for fumbling the future[3]—Xerox's PARC research lab famously invented the personal computer and graphical user interface, only to watch from the sidelines as Apple, IBM, and Microsoft built the PC business—but a brief tour of the AltaVista story shows that Xerox is certainly not alone.

The mighty rise and fall with spectacular regularity in this business, and the pace of boom and bust only increased as the Internet took root in the mid-1990s. Yet AltaVista is remarkable for a number of reasons. To borrow from the present, AltaVista was the Google of its era. In 1996, it was arguably the best and most-loved brand on the Web. It presaged many of the current innovations and opportunities in search, from automatic language translation to audio and video search to clustering of results. And as a business AltaVista attempted—and failed—to go public three times in three short years under three different owners. Possibly most instructive, AltaVista was the product of a company that was an extraordinary success in its original business but ultimately failed because of hidebound management unwilling to drive by anything other than the rearview mirror.

Monier Paints the Web

Regardless of the scuffle over its creation, it was Louis Monier who took AltaVista from concept to executable code. He came to the Western Lab from Xerox PARC, and the irony is not lost on him. "One reorg too many," is how Monier couches his decision to leave PARC (interestingly, the CEO of Google, Eric Schmidt, is also a Xerox alumnus).

"I've always been interested in big, nasty problems," Monier told me. Search provided one of the nastiest. Not only do the numbers scale to the near infinite, there was a very real need for good search in 1994. "Search engines at the time were just terrible," Monier recalls. "Yahoo was a great catalog, but it had no search. So I set about to work on the crawl."

As discussed in Chapter 2, traditional search engines have at their core three components. First is the crawl (or spider), which gathers every possible page on the Web. Second is the index, the massive database created by that crawl. And the third comprises the user interface and search software, which take the index and make it available in an intelligent fashion to the end user.

In 1994–1995, the Web was still new, and no one really had any idea how big it was or how quickly it was growing. But everyone in the industry knew it was huge, and growing on a scale that made engineers and mathematicians fibrillate—the numbers multiplied over the near term to a nearly infinite horizon. For Monier, the idea of creating an engine that might be considered the last word on the size of the Internet felt like a shot at immortality.

Nearly a dozen search engines already existed, but they fell short for one reason or another. Some had terrible user interfaces or lacked powerful query languages. Most indexed only URLs, not the entire content base of a Web site. Using the Alpha processor's considerable power, Monier constructed a new kind of crawler. This was critical to Monier's goal of completeness—he wanted to create an index of the entire Web, not just of URLs.

A crawler works in a linear fashion, discovering link after link and storing each page it finds along the way. Limited to one chain of discovery, a crawler would never find the entire Web—there are simply too many links, and too much time is needed to uncover them all. By the time it finished, the Web would have already increased significantly in size, and the task would have become impossible.

Solving for this scale required multiple crawlers that worked in parallel, building the Web index together. Thanks to the Alpha's 64-bit memory capability, Monier was able to set a thousand crawlers loose at once, an unprecedented feat. What they brought back was the closest thing to a complete index the young Web had ever seen—10 million documents comprising billions of words.

Monier hacked up an interface to the new index and tested it for two months internally at DEC. Everyone who used it loved it. But when Monier sought approval to release his engine to the public, DEC execs scratched their heads. What good was a search engine when it came to selling hardware?

Monier was nothing if not capable when it came to pressing DEC's buttons: he promised that AltaVista would generate good publicity, something the company sorely lacked. On December 15,

1995, Monier lifted DEC's firewall and gave the public access to altavista.digital.com, which by then had indexed more than 16 million documents.

But We're in the Minicomputer Business!

The year 1995 was a major one for search, with more than a dozen new companies formed, but it was a terrible time to be in the minicomputer business. Just five years earlier, DEC was near the height of its power, with $14 billion in revenue and more than 130,000 employees. Its VAX line of minicomputers powered a huge percentage of corporate data centers—the very data centers that would, by the late 1990s, be a driving force of the Internet revolution.

But by the mid-1990s, the company was bleeding money: $2 billion a year. It overexpanded in the late-1980s boom, and was ill prepared to compete in the brave new world of PC-based servers and desktops (though it did try). DEC was in the minicomputer business, and its executives were ill suited to compete with the likes of Compaq or Dell.

In those waning days of DEC's power, company brass reeled from one strategy to the next, cutting tens of thousands of jobs, launching a software division one day and new PC lines the next. Ultimately the company latched onto the Internet as a potential salvation—competitors SGI and Sun were selling high-powered Web servers, and perhaps DEC could as well. To drive the demand, DEC focused its software division on Internet connectivity and security tools. This was a classic example of corporate myopia—executives at DEC were attempting to fit a sleek new computing paradigm into their dowdy old product line. They hoped the Net would force customers to buy minicomputers. Instead, the Internet heralded and strengthened the personal computer revolution—the very trend responsible for killing off DEC's old line of business.

And yet DEC could lay claim to the mantle of Internet pioneer. If DEC was ever to strike Internet gold, it'd be at the Western lab. It held what was at the time one of the largest repositories of Internet content in the world—a machine called the Gatekeeper. Gatekeeper was a massive computer with unheard-of amounts of storage and an extremely fat pipe into the early Internet.

Researchers had created Gatekeeper in the "spirit of the public good," recalls Brian Reid. It served as a sort of public space where anybody could store and share any digital file, and thousands of early Internet technical innovators did just that. DEC may have been flailing in the corporate minicomputer market, but in the nascent Internet industry, it had serious street cred.

The First Google

Monier shakes his head as he recalls what happened after AltaVista launched. He couldn't have been more right about the publicity AltaVista would generate, but "we were too successful for our own good," he rues. With no marketing and no formal announcement, AltaVista garnered nearly 300,000 visits on its first day alone. Within a year, the site had served more than 4 billion queries. Four *billion*—nearly as many queries as people on Earth. This was truly a very big deal.

Monier's bosses at DEC were overjoyed with the press AltaVista was receiving. "The executive team was stunned," Monier recalls. "They still didn't understand the opportunity, but they loved the publicity."

They loved it so much, in fact, that in one meeting, a DEC PR executive created a fat roll of all of the site's press clippings and, to much celebration, ceremoniously unrolled the trophy across a boardroom table. But Monier remains dark as he describes what should have been a triumphant launch. "These people, they were used to hardware products," he says, reserving particular contempt for the word "hardware." "Remember, this is the same company that delayed the Alpha for eighteen months because they didn't like anything that wasn't a minicomputer. So when the press requests starting pouring in, they media-trained me in a hurry and came up with the 'demo for new hardware' justification."

While more diplomatic than Monier, Ilene Lang, the first CEO of AltaVista, won't take the bait when I offer her a chance to counter Monier's claims. Lang, who was hired away from a senior position at Lotus to run DEC's software group in 1995, joined just before AltaVista was slated to launch. "We knew this was a very big deal," Lang tells me. "This was about more than showing off the power of DEC hardware."

After seeing AltaVista and a few other Internet-related goodies at the Western Lab, Lang quickly reorganized her nascent software group into an Internet play, with AltaVista as its crown jewel. But Lang and Monier were frustrated by DEC's internal politics—the red-hot AltaVista couldn't get the resources, the attention, or, most important, the decisions it needed to move as quickly as its dot-com competition. DEC loved its new creation, but had no idea how to manage it.

And as the demand increased on the AltaVista site, Lang and Monier struggled to keep up. "Nobody would volunteer funds to grow the business," Monier recalls. Of course, he had all the hardware he needed, but search cannot live on hardware alone.

Making matters worse, Monier and Lang were not exactly drinking buddies. While Lang believed her division should sell a wide variety of Internet software solutions—security, search, e-mail, and the like—Monier was maniacally focused only on search. Of his Internet software business unit, he claims, "two hundred people were selling crap, and six of us were doing AltaVista."

"Louis had a one-track mind," Lang recalls, with a diplomatic tone. "He was often difficult to work with, and had no respect for the software business."

Unfortunately, in 1996, it was impossible to create a pure play in search that was economically viable. The market was still too immature—robust business models were years from fruition. Paid search innovator GoTo.com didn't exist, and "google" still meant 1 followed by 100 zeroes. To her credit, Lang didn't force Monier to turn AltaVista into a portal. As long as Lang was running the company, AltaVista remained remarkably focused on search, and it forged ahead as an advertising/sponsorship-based business, albeit a modest one compared with the revenues of its parent company. To justify her new division, Lang created a line of AltaVista-powered Internet software applications targeted at the enterprise customers DEC had sold to for decades. It made sense given the circumstances in which she found herself.

For DEC, of course, AltaVista really was a means to sell more hardware. The irony of this should not be lost to history. According to Gordon Bell, an Internet pioneer and early VP of R&D at DEC who now works as a researcher at Microsoft, DEC was the very first company to establish a dot-com—dec.com in 1985.

Leveraging AltaVista's heat and facing DEC's reluctance to invest its own money, Lang managed to convince the DEC board that AltaVista needed capital and public currency to grow. In August 1996, DEC decided to spin AltaVista off as a public company. For Monier and his team, the fruits of their long labors were nearly within reach.

But before AltaVista was able to make its public debut, DEC entered the throes of yet another reorganization. This time DEC decided to become a "solutions" company and abandon the business-unit-driven approach that had allowed AltaVista at least a semblance of independence. Lang and Monier fought to protect AltaVista from its flailing parent, but a mammal chained to a dinosaur more likely than not will get trampled. AltaVista was disbanded as a business unit and tossed into DEC's new structure as part of the marketing division. "Everyone else was being dismantled," Lang recalls being told, "so you should be, too." Frustrated and without a real role, she left shortly thereafter.

Monier stayed on, however, out of both love for his creation and perhaps a bit of madness. He believed that in the end AltaVista would prevail. "I should have left," he told me. "But I wanted to keep our principles intact." In other words, Monier wanted to make sure AltaVista stayed pure—the best search on the Web. "A pencil," Monier called it—a tool that did one thing very, very well. If that sounds familiar, it should—it's exactly the approach that catapulted Google to the top of the heap four years later.

By 1997, AltaVista was truly king of search. Serving more than 25 million queries a day and on track to make $50 million in sponsorship revenue, the company was in a three-way heat with Yahoo and AOL as the most important destination on the Web. And in an ironic foreshadowing of Google's role just a few years later, AltaVista captured the plum assignment of serving Yahoo's organic search results (Yahoo, at this point, was still convinced that its directory and portal services were the most important portion of its business).

Then the gunslingers showed up.

The Compaq Portal

In January 1998, DEC finally threw in the towel as an independent company, and agreed to a $9.6 billion acquisition by Compaq. AltaVista became a unit of a Houston-based personal computer giant with absolutely no knowledge of the consumer Internet. According to Monier, AltaVista carried almost no book value in the transaction, though in press interviews Compaq CEO Eckhardt Pfeiffer did promise to expand his newly acquired Internet company.

That turned out to be an understatement. While DEC's brand of parenting ran toward benign neglect with the occasionally irritating habit of taking credit for its progeny's accomplishments, Compaq quickly saw in AltaVista a chance to cash in on the burgeoning Internet bubble. It had one of the hottest brands on the Net, and as Monier puts it, "An entire division of Compaq thought they were going to get rich by taking over AltaVista.

"In the Houston headquarters there were literally signs that asked people to check their guns at the door," Monier recalls. "They got here and went berserk."

Rod Schrock, a Compaq executive widely considered to be a Pfeiffer protégé ("complete megalomaniac," mutters Monier), was given charge of AltaVista, and he immediately hired a battery of East Coast consultants to lay out his strategy for the company. The consultants told Schrock what he already wanted to hear: AltaVista had the brand and the technology to beat the portals at their own game. Build AltaVista into a Excite and Yahoo killer, and you will certainly be able to take the sucker public.

Within a year Schrock had turned AltaVista into a Yahoo clone, with e-mail, directories, comparison shopping, topic boards, and scads of advertising on the front page. He went on an acquisition spree, spending more than a billion dollars to purchase Zip2, a "portal services company"; Shopping.com; and Raging Bull, a financial site, among others. He dusted off AltaVista's first IPO filing and laid in plans for a second attempt at milking the public markets.

But Monier had finally had enough. In the spring of 1999, he quit, taking thirty members of his team with him. He held no stock, and took only his memories, his experience, and the license plates on his car, the plates he still uses to this day: ALTVSTA. "I'd rather do something interesting than something boring and get rich," he later said.

Schrock charged on, but before he could execute his plans for an IPO, Compaq decided to cash out on its Internet asset without the fuss of an IPO. It sold AltaVista to CMGI, a high-flying Internet holding company, for $2.3 billion (mostly in soon-to-be-worthless stock) in June 1999.

CMGI relaunched AltaVista that fall with a $100 million ad blitz. The company's strategy was not particularly innovative: build the best portal, then take it public. In December, CMGI filed paperwork for yet another AltaVista IPO, and scheduled it for the following April. But the NASDAQ index peaked on March 10, 2000. Just before the offering, the NASDAQ began its historic slide, losing nearly 35 percent of its value in less than a month. The bubble had burst.

CMGI shelved the IPO plans. Hoping the worst was over, in January 2001, the company filed again. By now, however, the markets were having none of it—the offering was pulled for a third and final time. Ever the child of wayward parents, AltaVista watched as the stock of its parent company, CMGI, lost more than 90 percent of its value. The once glorified engine limped along with very little support and a waning user base until what was left of the company was sold to paid search innovator Overture Services, Inc., in 2003. The price: $140 million.

Overture itself was later sold to Yahoo, which restored AltaVista to its original look: a search box, a blinking cursor, and scads of white space. But by then, AltaVista no longer was at the table. Monier, creator of the first Google, is now working at eBay, helping that commerce giant redesign—what else?—its approach to search.

Rise of the Big Guys

By 1995, several other major Web destinations had formed, including Lycos, which began life as a Carnegie Mellon University (CMU) project, as well as Yahoo and Excite.

Lycos was created in May 1994 by CMU's Dr. Michael Mauldin, working under a grant from the Defense Advanced Research Projects Agency (DARPA). The name was derived from Lycosidae, the Latin word for the wolf spider family, whose members actively seek their prey rather than catching it in a web. Like its predecessors, Lycos deployed a spiderlike crawler to index the Web, but it used more sophisticated mathematical algorithms to determine the meaning of a page and answer user queries. And it became the first major engine to use links to a Web site as the basis of relevance—the underlying basis for Google's current success.

The cornerstone of Lycos's technique was analysis of anchor text, or the descriptions of outbound links on a Web page, to get a better idea of the meaning of the existing page. A link such as "click here for more information about aspirin" at the bottom of a page might lend some context. It also used outbound links on a page to build and promote a bigger index, even if it hadn't crawled those pages. In another novel approach, Lycos introduced Web page summaries in search results, rather than a simple list of links. Previously, engines like WebCrawler displayed only the title and ranking of the page so that more results could be displayed at once.

Based in Waltham, Massachusetts, Lycos was the only East Coast operation in a sea of Silicon Valley start-ups. In June 1995, Carnegie Mellon sold 80 percent ownership of the Lycos technology to Mauldin and founding executive Bob Davis for $2 million. Backed by the university and CMGI's @Ventures (the venture arm of the company that owned and then sold AltaVista), the company got caught up in the dot-com frenzy. Just ten months after it was founded, Lycos went public and proceeded to make the same mistakes AltaVista would—snapping up several companies over the next few years, including home-page publisher Tripod and Wired Digital,[4] which owned rival search site HotBot.

For a short period in 1999, Lycos became the most popular online destination in the world. In May 2000, at the height of the bubble, Lycos was sold to Terra, a Spanish telecom giant, for $12.5 billion. Four years later, Terra sold Lycos to a South Korean company for about $100 million. (The AltaVista story has many cousins.) Today Lycos remains a top-twenty destination, but it has struggled to regain its past glories in light of the extraordinary success of Google.

Excite

If Vinod Khosla had had his way back in 1996, Excite might have avoided a similar fate. The legendary partner at Valley venture firm Kleiner Perkins Caufield & Byers (the same firm that later funded Google) was an early backer of Excite, and tried mightily to get its young founders to buy Google when it was still a research project.

The acquisition did not come to pass, and Excite eventually failed, though not before making its own particular mark on the history of search.

Founded in 1994 by six Stanford University alumni, all tight friends since their freshman year in the dorms, Excite began life under the name Architext. The company's original goal was to create search technology for large databases within corporations, but Khosla encouraged the company to focus on the consumer Web, going so far as to personally purchase for the entrepreneurs a computer large enough to hold the site's Web index. In the end, Khosla funded Excite with $1.5 million in seed money; another $250,000 came from Geoff Yang, another respected Valley VC.

Khosla cast a veteran's jaundiced eye on the early days of search. "Yahoo was running a directory, and we were running a text search paradigm—text search was much more interesting," he recalls. "I tried to get Yahoo and Excite to merge, but [Yahoo founders] Jerry [Yang] and [David] Filo said no." Khosla then looked east, toward Lycos, which at that point was still a CMU research project. "I tried to get them to buy Lycos for $1 million but . . ." Khosla rolls his eyes, frustrated by the memory of dealing with Excite's founders, fresh out of college—kids, basically, who thought they knew all the answers. "Because of their early success, they were closed-minded and a bit arrogant," Khosla recalls. "Nothing deceives like success," Excite cofounder Joe Kraus acknowledges.

The kids brought in adult supervision by 1995, hiring George Bell, a magazine executive, as CEO. "We were late to the market," recalls Brett Bullington, an early Excite executive. "Yahoo was already doing a million pages a day when we were founded."

Excite debuted in the fall of 1995 with a Web directory and full-text search engine with the tagline "twice the power of the competition." Excite was the first search engine to transcend classic keyword-based searching with technology that grouped Web pages by their underlying concepts. It used statistical analysis of word relationships on the page to deliver fine-tuned results to Web surfers.

But Excite was a perennial second place to top Web property Yahoo, and the fact that both companies started at Stanford only intensified the competition. To grow, Excite needed more capital and more traffic, and it turned to the same place nearly all portals did—the public markets. The company went public in April 1996 with a valuation of $177 million, and then began an acquisition and feature-building tear. It bought search rival Magellan for about $18 million, and shortly after acquired WebCrawler for $4.3 million from AOL.

But Excite did more than buy companies; it also pioneered key features now taken for granted on the Web. One of its most persistent innovations was personalization—MyExcite was among the first services to allow users to create custom Web pages with news, business information, and regional weather reports. And in the summer of 1997, Excite became the first of the major portals to offer free e-mail—a move that rivals Yahoo and Lycos would make that October. (Google finally capitulated and announced Gmail—its version of free e-mail—seven years later.)

Intent on winning the portal wars, Excite bid for ICQ, an extremely popular (but at the time revenue-free) Internet chat service founded in Israel. But Excite didn't have the cash to make the deal, and AOL ended up with the prize. "It was clear we had to bulk up or we had to partner," recalls Bullington. "Yahoo's stock was trading at a major premium to ours."

Excite played a central role in what might be called the great search scrum of 1998. Nearly every major search company was in play, and there was no more determined deal maker than Excite, which held mergers and acquisitions discussions with Yahoo, Google, AOL, Microsoft, and Lycos. According to both Khosla and Bullington, Excite was extremely close to closing a deal with Yahoo—the combined company would have owned a commanding lead in Web traffic—when another bidder came knocking on Excite's door.

When @Home, a broadband company owned by several major

cable companies, made a richer offer to combine Excite with its @Home broadband Internet service, the Excite team felt compelled to accept it. First, it was more money, but more important, the @Home team promised to aggressively take on AOL and Yahoo, beating them both with a combination of high bandwidth access and high-value media content.

Well, that was the idea, anyway. In the end, @Home was to Excite what Compaq was to AltaVista—a heady combination that turned out quite badly. @Home had a complicated relationship with AT&T, which had just purchased TCI, the largest shareholder in @Home. "AT&T changed its strategy and started playing politics," Khosla recalls. "They decided to get out of the media business. That killed Excite. In retrospect, we should have done the Yahoo deal."

Excite ended up in a very messy Chapter 11 proceeding, but its assets live on, sold for pennies on the dollar to Interactive Search Holdings (ISH), a small search holding firm, in 2002. ISH, in turn, was sold to Ask Jeeves, the perennial third-place search player, in March 2004. (Ask Jeeves later became acquisition fodder for Barry Diller's InterActiveCorp in early 2005.)

"All the portals suffered from the classic business mistake of veering from their core mission," summarizes Kraus. "Unbeknownst to them all, there was a giant vacuum left in search." That vacuum, of course, would soon be filled by Google. But Google came to power aided by the titan of Internet portals, Yahoo.

Yahoo

This isn't the official story, but the truth is, Yahoo got its start when two bored PhD candidates at Stanford hacked together a system that helped them win a fantasy basketball league.

Jerry Yang and David Filo were both pursuing doctorates in electronic design automation, a once-hot field that had cooled by the fourth year of their doctoral work. "The prospects of finishing and getting on with life were pretty grim," Yang recalls. "The real story is that we were bored with our PhDs and we did everything we could to avoid writing our thesis."

In the early 1990s, Yang and Filo worked (or rather, avoided working) together in a temporary building on Stanford's campus. To compete effectively in the fantasy league, Filo hacked up an Internet crawler that pulled data from basketball sites via protocols like FTP and Gopher—at the time, the Mosaic browser had not burst onto the scene, and the World Wide Web was still an academic experiment. Filo then compiled the data—statistics on players' performance, trade information, and the like—and together the duo analyzed it to determine their picks. They ended up winning the league.

"That was the first crawler that I knew about," Yang recalls. "It was one of those things where you realize if you could figure out how to unify all those protocols out there, you'd have something."

In 1993, Mosaic, the first Web browser, launched, and Yang started obsessively surfing the Web, keeping lists of sites he found interesting. Filo took note of Yang's passion and wrote some software that helped automate the list and together they published it on the new Web medium. Yang had already created a home page, Akebono (named after a famous sumo wrestler), on his student machine, and by default that became the list's first home. Jerry and David's Guide to the World Wide Web, the first iteration of what would later become Yahoo, made its debut in late 1994.

Jerry and David's Guide became one of the earliest viral success stories of the nascent Web—it grew by word of mouth, first within the tight-knit community of Stanford graduate students, then quickly outward to the entire Web. Within the first thirty days, the site had logged visitors from thirty countries, a fact that still astounds Yahoo's founders. Initial traffic started in the thousands of visitors but quickly scaled to the point where Yang's machine was consumed by the demand—not such a bad development for a student looking to avoid doing actual work.

In 1995, Yang and Filo decided to get serious about their endeavor by giving it a more memorable name. Inspired by computer science acronyms that started with "YA"—for "yet another"—Yang and Filo pulled out a dictionary and started at "Y." When they got to "Yahoo" they knew they had a winner.[5] Not only did they like the self-effacing double entendre—the dictionary defined the term as "a rude, unsophisticated, uncouth person"—but the word also lent itself to reverse engineering by way of acronym: Yet Another Hierarchical Officious Oracle.

Hierarchy was important to the early site. As it grew and the number of links increased, Yang and Filo adopted a directory approach to navigation—sorting links into categories like Arts, Science, Business, and so on. Subcategories blossomed underneath, and by the end of 1994, the site had ballooned to thousands of links. Traffic doubled every month, and it was clear the pair had a hit on their hands.

A success story like that was bound to get attention, particularly given that the Internet was generating buzz among the Valley's venture-capital community. Nowhere was that community more plugged in than at Stanford. Yang and Filo began to field calls from interested investors and they realized they needed to come up with a business model. "We knew we needed to get the site off of Stanford servers," Filo continues. That meant paying hosting and bandwidth costs, and that meant the founders needed cash.

"I think the first time we realized that, hey, there might be some money here," Filo says with a wry smile, "was when somebody approached us and wanted to publish our directory on a CD." Yang and Filo passed on that idea, but they began puzzling over the new medium for hours on end, posting new links to their site between meetings where ideas like selling books on the Internet were discussed and discarded (Amazon's Jeff Bezos is still thanking them for that one). In the beginning, Filo and Yang agree, they had no sense that the core driver in their new business—navigation—had any value at all.

"This only proves we're not the brightest guys in the world," Yang quips drily.

As Google's founders later discovered, Stanford's 6,200-acre patch of rolling California woodlands is the most productive incubator of technology companies the world has ever seen. Nestled between the silicon factories of Intel and Apple on one end and Sand Hill Road's venture capitalists on the other, Stanford is a place where students have always dreamed of starting their own companies or going to work for a pre-IPO start-up. And Stanford's computer science department, where Yang and Filo hung their hats, is perhaps the most prodigious start-up incubator of them all.

In such an environment, two bored doctoral candidates who stumbled upon Internet gold had to be out of their minds not to start a company to mine it. Much as Page and Brin would do two years later, Filo and Yang began to talk to various companies about selling their project, but most had no interest. The VCs encouraged them to set out on their own, and in March 1995, they accepted $2 million from Sequoia Capital's Michael Moritz (who later also funded Google).

But the elusive business model had yet to be invented. In October 1994, HotWired, a Web content portal created by *Wired* magazine, had gone live with a new approach to revenue borrowed from its print cousin: advertising.[6] Filo and Yang took note, as did much of the Internet world, and by late 1995, Yahoo had adopted the standard. Yahoo, which now counts its advertisers in the hundreds of thousands, first went live with banners from just five.

Yahoo had plenty of competition in the early days—by this time, there were literally dozens of sites that organized the Web, and AOL was gaining traction as well. But Yahoo's directory stood out—it organized the Web in a fashion that made sense to techies and first-time Web surfers alike. In the early days, "people got caught up in the directory versus search debate," Yang says, "but our approach was quality. How can technology give quality results?"

"Early on you couldn't put a search box in front of people and expect that they would know what to do," Filo adds. Most Web users were new to the experience; there were no preset habits attendant to surfing. A hierarchical approach simply made sense for a public trying to understand the wild and rather disorganized chaos of the early Web. As surfers moved from a stance of exploration ("What's out there?") to expectation ("I want to find something that I know is out there"), search as a navigational metaphor began to make more sense. In late 1995, Yahoo added search to its directory through a partnership with early search innovator Open Text. Later that year it switched to AltaVista.

Srinija Srinivasan, who joined Yahoo in 1995 as editor in chief, says, "The shift from exploration and discovery to the intent-based search of today was inconceivable. Now, we go online expecting everything we want to find will be there. That's a major shift."

Another reason Yahoo suceeded was its sense of fun—a characteristic that would come to define not only Yahoo, but nearly every Internet company seeking the fickle approval of the Web public. Yahoo pioneered some of the Web's earliest social mores—including, for example, links to competitors' sites in case a searcher could not find what he or she was looking for, and listing "what's hot" prominently on its home page, thereby driving extraordinary amounts of traffic to otherwise obscure sites.

Thanks to practices like these, the company captured the public's imagination early and often, garnering a slew of adoring press notices familiar to anyone watching Google's rise to prominence over the past few years.

Growing Up

Tim Koogle, Yahoo's first CEO, knew he was onto something when he met Yang and Filo in the summer of 1995. "When I met Jerry and Dave, I saw great guys who were clearly in need of adult supervision," Koogle tells me. "These were guys who were doing it for the right reason—passion—who had spent no marketing money but had a huge user base. Clearly, there was value being created."

Koogle focused the company on its core value proposition, that of navigation. "The Net is all about connection, but you can't connect people without good navigation," Koogle says. "We sat in the middle, connecting people."

The Yahoo team quickly realized the value of its users' clickstreams. "People came to our servers and they'd leave tracks," Koogle says. "We could see every day exactly what people thought was important on the Internet."

Leveraging that insight, Koogle and his team built out Yahoo's now sprawling business, launching Yahoo Finance, Yahooligans (a kids' site), and many other popular divisions.

Yahoo's popularity brought competition, and a constant tension between partnership and all-out business warfare. In 1995, according to an executive familiar with the company's inner workings, Ted Leonsis of AOL placed a call to Jerry Yang and bluntly told him that if Yahoo didn't sell to AOL for the set price of $8 million, AOL would kill the company within the year.

Yahoo's founders knew they needed help—within months of closing their financing, they had hired a team that complemented their strengths and addressed their weakness. Both Filo and Yang readily admit their lack of business expertise at the time, and welcomed the experience of Koogle, who was a former Motorola executive. Koogle ran the business, Yang focused on product, and Filo tended to the company's ever-growing technology infrastructure. Again, if this sounds familiar, it's because it's pretty much the exact same route Google would take a few years later.

In the mid-1990s, "running the business" meant wrangling with partners as much as anything else. With Excite, Netscape, AOL, Lycos, and scores of lesser entrants in the game, Koogle spent much of his time either fighting off acquisition attempts or proffering them. And then there was the complicated maze of traffic deals that stitched all the major portals to each other.

At the center of that web was Netscape. Because first-time users of its Web browser came to Netscape's home page, the company quickly became the most significant source of traffic on the Internet. Yahoo was awarded a top link on Netscape's site—a link that brought even more traffic and business Yahoo's way. In fact, for a while Netscape even hosted Yahoo's service on the Netscape site. "I had to put an end to that," Koogle says with a laugh, adding that it doesn't make sense to have your business in the hands of a potential competitor.

But while Netscape was the lord of traffic, it decided to make its business in software, with ancillary revenue in media. Linking to Yahoo was an afterthought, at least at first. Over time Netscape realized the power it wielded and sold its links to the highest bidder. By then, however, Yahoo was firmly established as one of the most popular destinations on the Web.

As the Web expanded and users' habits changed, Yahoo added more traditional search functionality to the site. But until 2003, Yahoo treated search as a partner-driven service. After Open Text and AltaVista, Yahoo moved on to Inktomi and ultimately Google.[7]

"We had to make a business decision about search," Koogle says, echoing similar comments from Yang and Filo. "Search as a standalone service was very capital intensive—so much storage and bandwidth. The economics had not yet emerged to justify the investment."

Koogle is right—search was and continues to be an extremely costly service to get right. The portals' fixation on traffic, and their neglect of search, had left a huge opening for someone to make a better mousetrap. Concerns about economics or business models didn't stop two more Stanford PhD candidates—Larry Page and Sergey Brin—from trying to reinvert search. Once they did, the world did indeed beat a path to their door.

Chapter 4
Google Is Born

Of all the frictional resistance, the one that most retards human movement is ignorance.

—Nikola Tesla

"If Edison had a needle to find in a haystack, he would proceed at once with the diligence of the bee to examine straw after straw until he found the object of his search. . . .

I was a sorry witness of such doings, knowing that a little theory and calculation would have saved him ninety per cent of his labor."

—Nikola Tesla,
as quoted in the *New York Times,* 1931

Heirs to Tesla

Larry Page always wanted to be an inventor. When he was twelve Page read a biography of Nikola Tesla, one of history's most prodigious inventors. Tesla discovered or developed the foundational technologies for an astonishing array of innovations, from wireless communication and X rays to solar cells and the modern power grid. But despite his extraordinary invention, Tesla remains a minor figure—in particular when compared with Thomas Edison, a man

Tesla worked for, fought with, and competed against for much of his career.

The twelve-year-old Page was struck by this fact: regardless of how brilliant and world-changing Tesla's work had been, the inventor received little long-term fame or fortune for his efforts.

Twenty years later, a pensive, distant look spreads across Page's features when he relates Tesla's story. For most of his life Tesla struggled to support his research, Page tells me. "He had all these problems commercializing his work. It's a very sad story. I realized Tesla was the greatest inventor, but he didn't accomplish as much as he should have. I realized I wanted to invent things, but I also wanted to change the world. I wanted to get them out there, get them into people's hands so they can use them, because that's what really matters."

It's fair to say that Page and his partner, Sergey Brin, have managed to avoid Tesla's fate. They've gotten their inventions into the hands of hundreds of millions of people. Along the way, they've made thousands of people very rich, improved the businesses of hundreds of thousands of merchants, and fundamentally changed the relationship between humanity and knowledge. In the process, Page and Brin have become fabulously wealthy and movie-star famous. And it did not take them a lifetime to do so. It took as long as the average doctorate in computer science—five years, give or take.

"I had decided I was either going to be a professor or start a company," says Page, when I ask him to recall his goals at the start of his graduate work in computer science in Northern California. "I was really excited to get into Stanford. There wasn't any better place to go for that kind of aspiration. I always wanted to go to Silicon Valley."

Page is not a person who does things on a whim. He speaks with the slightly pinched and oddly inflected accent of the supersmart, a rather nerdy tone that is sometimes mistaken as Eastern European. In fact, he's from Michigan; it's his partner, Brin, who hails from Russia. Old friends remember Page as intelligent, ambitious, and nearly obsessed with efficiency. As an undergraduate at the University of Michigan, while president of the engineering honor society, he spearheaded a quixotic effort to build a monorail from one side of the campus to another because it seemed efficient (it was never built). In this manner, Page reminds many of another famously efficient founder: Bill Gates, founder and chairman of Microsoft. The comparison has followed Page throughout his tender career, and not simply because Page shares a tic or two with the richest man in the world.[1] In Google, many see a company that someday may supplant Microsoft as the most important—and most profitable—corporation ever created.

It Began with an Argument

Larry Page and Sergey Brin both knew what they were getting into when they accepted admission into Stanford University's graduate school of computer science. Stanford's elite program is known worldwide for its heady mix of academic excellence and corporate lucre. Students don't come to Stanford just for the training. They come for the dream: to start a company, grow rich, make their mark on the history of technology, and maybe change the world. This is the university, after all, that spawned Hewlett-Packard, Silicon Graphics, Yahoo, and Excite, to name just a few. Most members of the computer science faculty have started, run, sold, and/or advised Valley-based companies. So to say that starting a company was on Larry and Sergey's minds when they showed up at Stanford is to understate the case.

Larry first met Sergey in the summer of 1995, before he had decided to accept Stanford's offer of admission. Like most schools, Stanford invites potential recruits to the campus for a tour. But it wasn't on the pastoral campus that Page met Brin—it was on the streets of San Francisco. Brin, a second-year student known to be gregarious, had signed up to be a student guide of sorts. His role that day was to show a group of prospective first-years around the City by the Bay.

Page ended up in Brin's group, but it wasn't exactly love at first sight. "Sergey is pretty social; he likes meeting people," Page recalls, contrasting that quality with his own reticence. "I thought he was pretty obnoxious. He had really strong opinions about things, and I guess I did, too."

"We both found each other obnoxious," Brin counters when I tell him of Page's response. "But we say it a little bit jokingly. Obviously we spent a lot of time talking to each other, so there was something there. We had a kind of bantering thing going."

Walking up and down the city's fabled hills that day, the two argued incessantly, debating the value of various approaches to urban planning, among other things. Even if they weren't sure they liked each other yet, they were drawn together—two swords sharpening each other. Page accepted the offer from Stanford.

When Page showed up at Stanford for his first year, he selected as his adviser Terry Winograd, a pioneer in human-computer interaction (HCI). Page began searching for a topic that might prove fruitful for his doctoral thesis. It was an important decision. A dissertation can frame one's entire academic career, as Page had learned from his academic father, a computer science professor at Michigan State. He kicked around ten or so intriguing ideas, but found himself drawn to the burgeoning World Wide Web. With Winograd's urging, he decided to focus his attention there.

Page didn't land on the idea of Web-based search at the outset; far from it. Despite the fact that Stanford alumni were getting rich starting Internet companies, Page found the Web interesting primarily for its mathematical characteristics. Each computer was a node, and each link on a Web page was a connection between nodes—a classic graph structure. "Computer scientists love graphs," Page tells me, referring to the mathematical definition of the term.[2] The World Wide Web, Page theorized, may have been the largest graph ever created, and it was growing at a breakneck pace. One could reasonably argue that many useful insights lurked in its vertices, awaiting discovery by inquiring graduate students. Winograd agreed, and Page set about pondering the link structure of the Web.

Citations and Back Rubs

It proved a fruitful course of study. Page noticed that while it was trivial to follow links *from* one page to another, it was nontrivial to discover links *back*. In other words, when you looked at a given Web page, you had no idea what pages were linking back to it. This bothered Page. He thought it would be very useful to know who was linking to whom. After all, very important people might be linking to you—if so, wouldn't you want to know that? Or perhaps people were linking to you with malicious intent. What if one of the most visited sites on the Web had a link to your page that said, "This is the most godawful site on the Internet"? If Page could create a tool that allowed sites to easily discover and declare their backlinks, the Web would become far more interesting.

Why? To fully understand the answer to that question, a minor detour into the world of academic publishing is in order. Its Byzantine rigors are not for the fainthearted, but a few concepts deserve elucidation. For professors—particularly those in the hard sciences like mathematics or chemistry—nothing is as important as getting published. Published papers become an academic's calling card, a living résumé. The papers also determine tenure, that is, job security for life.

Academic publishing depends on peer review, the critical evaluation of a work by peers in the author's chosen field. Peer-reviewed journals are publications edited by experts who know how to critically assess a particular work and determine its academic importance. It is the goal of nearly all academics to have their papers published in peer-reviewed journals.

In addition to peer review, academic publishing turns on the idea of citation. There are many definitions of citation, but the library at the University of Massachusetts nails it: "A reference or listing of the key pieces of information about a work that make it possible to identify and locate it again." Academics build their papers on a carefully constructed foundation of citation: each paper reaches a conclusion by citing previously published papers as proof points that advance the author's argument.

Consider, for example, the citations in the following passage from "Authoritative Sources in a Hyperlinked Environment," a widely cited paper on search by Cornell University's Jon M. Kleinberg:

Bibliometrics [22] is the study of written documents and their citation structure. Research in bibliometrics has long been concerned with the use of citations to produce quantitative estimates of the importance and "impact" of individual scientific papers and journals, analogues of our notion of authority. In this sense, they are concerned with evaluating standing in a particular type of social network—that of papers or journals linked by citations.

The most well-known measure in this field is Garfield's impact factor [26], used to provide a numerical assessment of journals in Journal Citation Reports of the Institute for Scientific Information. Under the standard definition, the impact factor of a journal j in a given year is the average number of citations received by papers published in the previous two years of journal j [22]. Disregarding for now the question of whether two years is the appropriate period of measurement (see e.g. Egghe [21]), we observe that the impact factor is a ranking measure based fundamentally on a pure counting of the in-degrees of nodes in the network.

Pinski and Narin [45] proposed a more subtle citation-based measure of standing, stemming from the observation that not all citations are equally important. They argued that a journal is "influential" if, recursively, it is heavily cited by other influential journals. One can recognize a natural parallel between this and our self-referential construction of hubs and authorities; we will discuss the connections below.

In this passage, Kleinberg first defines a term (bibliometrics). He then cites the authority in the space (the legendary Eugene

Garfield, who is widely credited as the father of citation analysis), and proceeds to cite those who have built upon Garfield's work (Egghe, Pinski, Narin). Finally, Kleinberg puts forward his own conclusions, based on his theories of hubs and authorities.[3]

Not exactly beach reading, but academic publishing follows the principles of scientific inquiry, demonstrating clear paths to logical conclusions by citing the works of others. (If you can recall being chided by your high school English teacher for failing to properly organize your footnotes and bibliography, you'll know what I'm talking about.) The process of citing others confers their rank and authority upon you—a key concept that informs the way Google works.

The penultimate concept that is germane to our tour of academic publishing is that of annotation. In an academic setting, annotation is clearly defined: it refers to the practice of adding descriptive notations to citations. These days, it can include criticism or commentary: I'll cite this paper, but its author labored under false pretenses for most of his life. An annotation is a judgment about the cited paper.

Finally, while there's no academic term for it, academic publishing is driven by the concept of rank. Papers are judged not only on their original thinking and the rigor of their citations, but also by the number of papers they cite, the number of papers that subsequently cite them back, and the perceived importance of each citation. While this practice has led to citation inflation (long-winded, pointless citational throat-clearing) as well as citation log-rolling (I'll cite you if you cite me), it does provide a rough ranking mechanism for any given paper. Indeed, Garfield, among many others, has shown that a given paper's importance can be ascertained by noting how many other papers link to that paper through citation.

Academic publishing, then, is a flawed but useful system of peer review, incorporating citation and annotation as core concepts. The system produces a ranking methodology for published papers.

Fair enough. So what's the point?

Well, it was Tim Berners-Lee's desire to address the drawbacks of this system, via network technology and hypertext, that led him to create the World Wide Web.[4] And it was Larry Page and Sergey Brin's attempts to improve Berners-Lee's World Wide Web that led to Google. The needle that threads these efforts together is citation—the practice of pointing to other people's work in order to build up your own.

Which brings us back to the original research Page did on backlinks. He reasoned that the entire Web was loosely based on the premise of citation and annotation—after all, what was a link but a citation, and what was the text describing that link but annotation? If he could divine a method to count and qualify each backlink on the Web, as Page puts it, "the Web would become a more valuable place."

"In a sense," Page continues, "the Web is this: anyone can annotate anything very easily just by linking to it. But the early versions of hypertext had a tragic flaw—you couldn't follow links in the other direction. BackRub was about reversing that. It seemed kind of cool to gather all the links on the Web and reverse them."

Page hypothesized BackRub, as he called his project, as a system that would discover links on the Web, store them for analysis, then republish them in a way that made it possible for anyone to see who was linking to any given page on the Web. An ambitious idea on any scale, but Page didn't set out to make BackRub work on a small set of test pages. Instead, he thought big: why not solve the problem all at once, for *the entire World Wide Web*?

To undertake such a task requires a rather audacious bent. While Page was storing just the links—not the contents of the entire Web—he still had to *crawl* the entire Web to find those links. In 1995, such a feat was quite rare.[5]

At the time Page conceived of BackRub, the Web comprised an estimated 10 million documents, with an untold number of links between them. Page figured that it was somewhere in the range of 100 million. The number turned out to be much larger. And the longer Page waited to get started, the bigger the Web became. In the early days, the Web was growing at a rate of more than 2,000 percent a year. The computing resources required to crawl such a beast were well beyond the usual bounds of a student project. Somewhat unaware of what he was getting into, Page began building out his crawler.

The idea's complexity and scale lured Sergey Brin. Brin, a polymath who had jumped from project to project without settling on a thesis topic,[6] found the premise behind BackRub fascinating. "I talked to lots of research groups" around the school, Brin recalls, "and this was the most exciting project, both because it tackled the Web, which represents human knowledge, and because I liked Larry and the other two people who were working with us."

The two others working with Page and Brin were Scott Hassan and Alan Steremberg, graduate assistants who had been assigned to the project. (Each PhD candidate was assigned an assistant or two—usually a master's student looking to make a little extra money.) Hassan and Steremberg ended up separating from the project before Google really took off. But even those missing Beatles started successful Internet companies. Hassan went on to found eGroups.com with Larry's brother, Carl Page, and later sold it to Yahoo for more than $500 million. Steremberg had already launched The Weather Underground, a popular weather site, while an undergraduate at Michigan, and still runs it today.

The Audacity of Rank

Page told me that it had never been his intention to create a search engine—indeed, he and Brin had no idea what useful things the project might turn up. But in order to create BackRub, they had to crawl the web. In March 1996, Page pointed his crawler at just one page—his own home page at Stanford (most CS grad students had one)—and let it loose. The crawler worked outward from there. That's the beauty of the Web—no matter where you start, eventually you'll get just about everywhere else there is to go.

Crawling the entire Web to discover the sum of its links is a major undertaking, but simple crawling was not where BackRub's true innovation lay. Page was naturally aware of the concept of ranking in academic publishing, and he theorized that the structure of the Web's graph would reveal not just who was linking to whom, but more critically, the *importance* of who linked to whom, based on various attributes of the site that was doing the linking.[7] As noted earlier, those attributes—the anchor text around the link, for example—are also critical in determining rank and relevance.

If BackRub knew the importance of a site, it could give that site a relative ranking. For any given site, one could see not only who was linking to that site, but the *ranking* of those links as well. Certainly *that* would be useful, Page thought.

Being useful was an extremely important aspect of Page and Brin's research (and has become a mantra for all of Google's product development since). They hadn't yet decided that there was a company in BackRub, but the lessons of Tesla were never far from Page's mind. "My goals were to work on something that would be academically real and interesting," Page recalls. "But there is no reason if you are doing academic work to work on things that are impractical. There are plenty of interesting problems that are also practical. I wanted both, and I didn't think there was much of a trade-off to be made. I figured if I ended up building something that was going to potentially benefit a lot of people . . . then I would be open to commercializing it—so that I wouldn't be like Tesla."

Once Page and Brin had crawled the Web and stored a graph of its links, they needed to determine a ranking methodology. Inspired by citation analysis, Page theorized that a raw count of links to a page would be a useful guide to that page's rank. He also theorized that each link needed its own ranking, based on the link count of its originating page. But such an approach creates a difficult and recursive mathematical challenge—you not only have to count a particular page's links, you also have to count the links attached to the links. Very quickly, the math gets rather complicated.

Fortunately, Brin's prodigious gifts in mathematics could be applied to the problem. Brin, the Russian-born son of a NASA scientist (his mother) and a university math professor (his father), emigrated to the United States with his family at the age of six. By the time he was a middle-schooler in suburban Maryland, Brin was a recognized math prodigy. He dropped out of high school a year early to enroll at the University of Maryland, where his father taught. Once he graduated he immediately enrolled at Stanford, where his talents allowed him to goof off. The weather was so good, he told me, that he took mostly nonacademic classes—sailing, swimming, diving. He focused his intellectual energies on interesting projects rather than actual coursework.

Together, Page and Brin created a ranking system rewarding links that came from sources that were important, and penalizing those that did not. For example, many sites link to ibm.com. Those links might range from a business partner in the technology industry—Intel, perhaps—to a teenage programmer in suburban Illinois who linked to IBM because he just got a new computer for Christmas. How might an algorithm determine rank between these two citations? For a human observer, the business partner is a more important link, in terms of understanding IBM's place in the world. But how might an algorithm understand that fact?

Page and Brin's breakthrough was to create an algorithm—dubbed PageRank after Page—that manages to take into account both the number of links into a particular site, and the number of links into each of the linking sites. This mirrored the rough approach of academic citation counting, and as it turned out, it worked. In the IBM example above, let's assume that only a few sites linked to the teenager's site. Let's further assume the sites that link to the teenager's are similarly bereft of links. In contrast, thousands of sites link to Intel, and those sites, on average, also have thousands of sites linking to them. Under PageRank, the teenager's site would rank as less important than a site like Intel. In this example, then, Page and Brin's ranking methodology would judge Intel as more important than a suburban teenager—at least in relation to IBM.

This is a simplified view, to be sure, and Page and Brin had to correct for any number of mathematical cul-de-sacs, but the long and the short of it was this: more popular sites rose to the top of their annotation list, and less popular sites fell toward the bottom.

As they fiddled with the results returned by their work, Brin and Page realized they were onto something that might have implications for Internet search. In fact, the idea of applying BackRub's ranked page results to search was so natural, Page recalls, that it didn't even occur to them that they had made the leap. As it was, BackRub already worked like a search engine—you gave it a URL, and it gave you a list of backlinks ranked by importance. "We realized that we had a querying tool, a page ranking that was useful for a lot of things," Page recalled. "It gave you a good overall ranking of pages and ordering of follow-up pages."

Page and Brin quickly noticed that BackRub's results were superior to those of traditional search engines like AltaVista and Excite, which often returned irrelevant results. "We thought, *Why are they returning these results that are obviously not important?*" Page recalls. "They were only looking at text and not considering this other signal. Once you have it, it's pretty obvious that this signal is useful in search."

The signal—now better known as PageRank—became the foundation of Google's vaunted secret sauce. To test whether PageRank worked well in a search application, Brin and Page hacked together a BackRub search tool. It searched only the words in URL titles and applied PageRank to rank the results for relevance, but its results were so far superior to traditional search engines—which ranked mostly on keywords only—that Page and Brin knew they were onto something big.[8]

And not only was the engine good; Page and Brin realized it would scale as the Web scaled—PageRank worked by analyzing links, so the bigger the Web got, the better the engine would be. That fact inspired the founders to name their new engine Google, after googol, the term for the number 1 followed by 100 zeroes. They released the first version of Google on the Stanford Web site in August 1996.

Among a small set of Stanford insiders, Google was a hit. Energized, Brin and Page began improving the service, adding full-text search and more and more pages to the index. But search engines require an extraordinary amount of computing resources. Graduate students usually lack the money to buy new computers; Page and Brin were no exceptions. Instead they begged and borrowed Google into existence—a hard drive from the network lab, an idle CPU from the CS loading docks. Using Page's dorm room as a machine lab, they fashioned a computational Frankenstein from spare parts, then jacked the whole thing into Stanford's broadband campus network. After filling Page's room with equipment, the young students converted Brin's room into an office and programming center.

Hector Garcia-Molina, another faculty adviser on the project, lent the students a Sun Ultra, a powerful computer that Page recalls had ten times the memory of a typical PC. But even that was not nearly enough. Acquiring resources for their work became nearly a full-time job. "We begged hardware and connectivity from other groups—we had so much data to move around," says Page. "We had to ask them to open up their wiring closets, lend us their Ethernet cards."

An early e-mail from Larry Page to Terry Winograd illustrates the kind of problems the founders encountered. Dated July 15, 1996, it reads in part:

> I am almost out of disk space. I have downloaded about . . . 24 million unique URLs, and about 100 million links. . . . I think I will need about 8 gigs more to store everything. . . . Current retail prices are about $1000/4 gigs. . . . I have only about 15% of the pages but it seems very promising.

Owing to its size and scale, the project grew into something of a legend within the computer science department and the campus network administration offices. At one point the BackRub crawler consumed nearly half of Stanford's entire network bandwidth, an extraordinary fact considering that Stanford was one of the best-networked institutions on the planet. And on at least one occasion, the project brought down Stanford's Internet connection altogether. "We're lucky there were a lot of forward-looking people at Stanford," Page recalls. "They didn't hassle us too much about the resources we were using."

But the administrators at Stanford were hassled by many Web site owners, most of whom did not understand why Google's service was constantly requesting copies of their sites' pages. Back in 1996, it was nobody's goal to be indexed by a search engine; a request to download the entire content of a site was often seen as tantamount to trespass. A typical visitor to a Web site might click around a site, viewing a few pages here and there, then move along to the next site. But the BackRub crawler consumed a site entirely, indexing each page at the speed of light. Often sites were simply not designed to take such a load; they would buckle under BackRub's ravenous demands. Even if the site could withstand the crawler's request, the process felt like a violation of some unwritten rule of conduct, if not something more malicious.

Winograd tells the story of an online art museum that contacted Stanford after BackRub had indexed the museum's site. Because the crawler had requested every single page on the site, the museum was convinced that BackRub's true goal was to steal the images and text of the museum and re-create it somewhere else. The museum threatened to sue, but Winograd negotiated a truce. Complaints such as these eventually raised the eyebrows of Steve Hansen, the computer security officer for Stanford University. He e-mailed the entire Google project team in February 1997:

> Over the past six or seven months I have received numerous complaints from off-campus web sites regarding excessive and/or unauthorized web accesses originating from . . . the Computer Science Department. . . . Mr. Page . . . has done little to placate web site operators. . . . If research is to be done out on the Internet it must be done with much more care and supervision that has been evident with the BackRub project. If we do not apply effective self-policing in this area it may be that others will decide that we need policing from the outside.

Page apologized, went to a meeting with Hansen, and promised to do better. He posted a Web page explaining to the public that while Google did index the entire Web, it did not keep copies of every page. He also detailed how a Web site owner could request exclusion from the BackRub crawler's industrious requests. But spurred by yet another complaint in April 1998, Hansen again e-mailed Page:

> This is not the first, or even the second time this project has caused problems for another web server on the net. This sort of thing has cost these folks significant dollar losses. . . . [This] certainly doesn't do much for the reputation of the University or the Computer Science Department. I am also concerned about potential liability.

Page managed again to placate Hansen and the project continued apace. (Page was clearly impressed with Hansen's skills; he later hired him to run security for Google.)

But the complaints were not simply about BackRub's use (or abuse) of computing resources. Site owners were beginning to pay attention to the Google search service itself, in particular to how their sites ranked according to the nascent PageRank algorithm. Many were not pleased with the upstart search engine's seemingly blind judgment regarding their site. After all, this was the first time anyone had claimed to rank the inherent value of a Web site. As it still does to this day, such judgment evoked a powerful response.

"I am in a state of stupefication regarding your ranking process," read one typical complaint. "Since you created this search engine, you need to correct a hideous error which is both laughable yet hurtful to a webmaster. Please type the words 'Ulysses S. Grant' into your search engine and look at the results. My website, 'The Ulysses S. Grant Home Page' was voted 'The Best Civil War Website' in the February 1998 issue of *Civil War Time Illustrated*. . . . This website is graded as the premier website on any Civil War personality or battle. Bill Gates even wrote me a personal email praising it. . . . You . . . rank other, inferior (and some pathetic) sites higher. . . . This is an injustice of such magnitude that it begs explanation. I feel confident that if you take 5 minutes to look at my website you will rank it higher."

Page and Brin had clearly hit a nerve, not just with Civil War aficionados, but with every person who labored over a Web site. To many, unleashing a ranking system based on a bloodless algorithm felt like a supreme act of arrogance—who were these kids from Stanford, telling the world how we ranked? What did they know about the work and passion that went into *our* sites?

Well, in truth, Page and Brin made no claim to such knowledge. As these early complaints illustrate, the Google service made no pretensions of actually reading a particular site, or of understanding its content. It simply laid bare the often ugly truth of how well connected a site happened to be. No matter how great a site might look, or how many awards it might receive, if it was not linked to by other sites—ideally, sites that were themselves well linked—then, in Google's estimation, it didn't really exist. That cold, hard fact was hard for many to swallow.

A May 1998 e-mail from Winograd to Brin about the complaints foreshadowed the power Google would soon have over nearly every site on the Web:

> Long ago, Larry came to me and was eager to do research by putting a service into general use on the web. I was skeptical because it opens you up to random hassles, with the number of hassles proportional to the number of people affected by your service. We have now crossed that line, and are in the position where stopping the service will create a large number of complaints as well. But I guess that is just the cost of doing business!

While Page and Brin didn't know it at the time, their early ranking system was etching the traces of an entirely new ecology, an ecology shaped by millions of decisions and millions of Webmasters, each one of them wishing simply to rank better in the Google index.[9]

A Company Emerges

As Brin and Page continued experimenting with search, BackRub and its Google implementation were gaining buzz, both on the Stanford campus and within the cloistered world of academic Web research. One person who had heard of Page and Brin's work was the aforementioned Jon Kleinberg, then a researcher at IBM's Almaden center in San Jose, now a professor at Cornell. Kleinberg's hubs-and-authorities approach to ranking the Web is perhaps the second most famous approach to search after PageRank.[10]

Back in the summer of 1997, Kleinberg visited Page at Stanford to compare notes on search. Kleinberg had completed an early draft of his seminal "Authoritative Sources" paper, and Page showed him an early working version of Google running on the makeshift system he and Brin had cobbled together. Kleinberg encouraged Page to publish an academic paper on PageRank.

But in the course of his conversation with Kleinberg, Page told Kleinberg that he was wary of publishing. The reason? "He was concerned that someone might steal his ideas," Kleinberg told me. It was Page's Tesla conflict at work: on the one hand Page respected and participated in the academic tradition of sharing research through published papers, but he was also influenced by the more closed, defensive posture of a corporation protecting its intellectual property. With PageRank, "[Page] felt like he had the secret formula," Kleinberg told me. "It did seem a bit strange at the time."

Academic fame ultimately won out over the proprietary impulse. By the end of their conversation, the pair agreed to cite each other in their papers. In early 1998, Page submitted his first paper, an overview of the PageRank algorithm, to the Special Interest Group on Information Retrieval of the Association for Computing Machinery (SIGIR-ACM). But the paper was rejected. One peer reviewer wrote of the paper, "I found the overall presentation disjointed. . . . This needs to focus more on the IR issues and less on web analysis." Page nevertheless persevered, and the paper was ultimately published in conjunction with a Stanford digital libraries project.

Shortly thereafter Page and Brin published a paper on Google itself. That paper, "The Anatomy of a Large-Scale Hypertextual Web Search Engine," has become the most widely cited search-related publication in the world. Given the ultimate success of Google itself, it seems Page and Brin had their academic cake and got to eat it, too.

Back in the early years, Page and Brin weren't sure they wanted to go through the travails of starting and running a company. During Page's first year at Stanford, his father had died, and friends recall that Page viewed finishing his PhD as something of a tribute to his father's life. Given his own academic upbringing, Brin, too, was reluctant to leave the program. Brin recalls speaking with his adviser, who told him, "Look, if this Google thing pans out, then great. If not, you can return to graduate school and finish your thesis." Brin chuckled, then added: "I said, 'Yeah, OK, why not? I'll just give it a try.' "

Through Stanford, Page and Brin had access to an extraordinary network of Silicon Valley business intelligence, and by 1997, Page's brother Carl was already hard at work building eGroups. The consensus view held that there were already a gaggle of search-related businesses, all well funded and thriving. Yahoo, Excite, AltaVista, Infoseek, Wired Digital's HotBot: the list was long and growing. Page and Brin reasoned that the best course might be to license their new technology to another company.

The inventors faced a classic entrepreneurial dilemma: if they started a company, it could be crushed by larger, richer competitors. On the other hand, if the company took off and became best of breed, the upside would be huge. Yahoo, Excite, and others already had multi-hundred-million-dollar valuations. But taking them on was risky. Page and Brin chose a more conservative course. Better to license the technology to a major player, they reasoned, and avoid the risks of a start-up.

The first attempt to license Google's technology occurred very early in the project's life. Vinod Khosla, the well-connected partner at the venture capital firm Kleiner Perkins Caulfield & Byers, had learned of Google through his own Stanford connections. Impressed, he tried to persuade a company he had invested in—the newly public Excite—to acquire the technology and its creators' services. This incited a flurry of e-mail between Khosla, Page, Winograd, and Brin. Page set the price for Google at $1.6 million. Khosla said he thought he could persuade Excite to offer $750,000.

Reviewing these early e-mail exchanges, it's remarkable to see Page's incipient business savvy. He knew that Excite was in heated battle with the much larger Yahoo, and saw Google's technology as a key to Excite's gaining a competitive edge. Wasn't that worth bridging the difference between his price and Khosla's counter offer? "The market leader usually is at least five times as big as the number two," Page wrote to Khosla, a veteran deal maker. "[Google's] significantly improved search technology will help Excite gain and maintain market share."

Page also argued that there would be a significant cost to Excite should his technology end up elsewhere, but the Excite executives were unconvinced. Khosla visited the Excite campus to persuade the CEO, George Bell, to change his mind. (Bell, a seasoned publishing executive, constituted the "adult supervision" brought in by Excite's investors.) "Bell threw me out of the office," Khosla told me with a wan smile. "At least I tried."

Over the course of the next eighteen months, the young inventors gave demonstrations of Google to nearly every search company in the Valley, from Yahoo to Infoseek. They also showed their technology to several venture capitalists. Everyone found their technology interesting, but each sent the grad students on their way. "I told them to go pound sand," recalled Steve Kirsch, founder of the now-defunct portal Infoseek. Jerry Yang and David Filo, the founders of Yahoo, were more encouraging, but they, too, took a pass.

"They were becoming portals," Page recalls of the companies he visited. "We probably would have licensed it if someone gave us the money. . . . [But] they were not interested in search.

"They did have horoscopes, though," he adds drily.

Suffice it to say, search was not top of mind for most Internet executives in the late 1990s. Search was a commodity—a feature that was "good enough." And anyway, in the late 1990s the goal was not to send people away from your portal, as search did. It was to keep them there.

Rejected but not deterred, Brin and Page went back to Stanford and kept working on Google, which they kept up and running at google.stanford.edu. "We said to ourselves, 'We don't care,'" Page says. "'We'll work on it some more. Maybe it'll turn into a company, or maybe it'll just be great research.'"

But by the middle of 1998, the service was growing at a rate that reminded Page of his brother's eGroups business. "It was getting more and more searches, and from Carl's experience with eGroups, we learned that if you have something that's growing like that, it just keeps growing."

By late 1998, Google was serving more than ten thousand queries a day, and it was clear to Page and Brin that the service would quickly outgrow their ability to beg resources to support it. Starting a company became the only viable alternative. The founders turned to another faculty adviser, David Cheriton. Cheriton, who heads Stanford's Distributed Systems Group, was an old hand at company formation. He had founded Granite Systems, a developer of networking technology that was sold to Cisco Systems in 1996 for $220 million. Cheriton suggested that Brin and Page meet with Andy Bechtolsheim, a founder of Sun who was active in early-stage investments.

As Page recalls, Brin sent Bechtolsheim an e-mail late one night requesting a sit-down, and Bechtolsheim answered immediately. He suggested meeting the next morning at eight o'clock, an hour at which the graduate students were unaccustomed to giving demos. But they agreed to meet, on the porch of Cheriton's Palo Alto home, which Bechtolsheim passed on his way to work each day.

"David had a laptop on his porch in Palo Alto, with an Ethernet connection," Page recalls. "We did a demo, and Andy asked a lot of questions. [Then] he said: 'Well, I don't want to waste time. I'm sure it'll help you guys if I just write a check.' "

Page and Brin weren't ready for such an offer, but when Bechtolsheim went out to his car to get his checkbook, they pondered how much to ask for and at what valuation. When Bechtolsheim returned, they told him their suggested valuation. Page picks up the story: "We told him our valuation, and he said 'Oh, I don't think that's enough, I think it should be twice that much.' "

Brin and Page were stunned, but of course, they agreed, and Bechtolsheim asked who the check should be made out to. The founders hadn't settled on a name, so Bechtolsheim suggested Google Inc., after the service's name. They agreed, and minutes later, Page and Brin had a check for $100,000. If ever there was a reason to incorporate, this was it.

To celebrate, Brin and Page went to Burger King and had

breakfast. "We thought we should [eat] something that tasted really good, though it was really unhealthy," Page said. "And it was cheap. It seemed like the right combination of ways to celebrate the funding."

The Early Years

Page kept the check in his dorm room desk for several weeks, as the founders went about forming the company and setting up bank accounts. On September 7, 1998, Google Inc. was formally incorporated, with Page as CEO and Brin as president. When Brin and Page hired their first employee—fellow student Craig Silverstein—they realized they needed to find office space, as the three of them could no longer work out of Sergey's dorm room. They found a temporary answer in Susan Wojcicki, a friend of Sergey's girlfriend.

Wojcicki, a recently graduated MBA, had just purchased a five-bedroom house in Menlo Park, a suburb near the Stanford campus. She recalls being worried about covering her mortgage payments, and when Brin and Page offered to rent a spare room, she agreed. (It didn't hurt that Brin had become Wojcicki's first customer in an online dried fruit business she had recently started.) Google Inc.—all three employees—moved in the next day.

"They went to Costco and filled their car with food," Wojcicki recalls. Concerned about her privacy—Wojcicki was pregnant at the time—Wojcicki insisted that her new tenants enter their offices through the garage door. The newly minted entrepreneurs not only had seed capital; they could now lay claim to the most shopworn cliché in the Valley—a garage address.

As Google grew, so did its fame. The founders raised additional capital (nearly a million dollars) from various well-connected angel investors—typically wealthy Valley businesspeople. Adviser David Cheriton came in, as did Ram Shriram, a former Netscape executive who had launched and sold a business to Amazon, where he was working as VP of business development. Shriram became a part-time adviser to the founders and persuaded his CEO, Internet superstar Jeff Bezos, to invest as well.[11]

In the months that Google occupied Wojcicki's spare room, the company focused on honing its service and preparing for a larger round of financing. It was in this makeshift office that Google entertained its first major press coverage—from *Time* magazine, which later included Google in its year-end roundup of the "best cybertech of 1999." It was also during this period—October 1998, to be exact—that Google adviser Winograd received this e-mail from a manager at Netscape, which at the time was the largest and most important destination on the Web:

> Hi Terry,
> Bunch of us here at Netscape have been playing with Google. There is significant interest in potentially using Google or a derivative as a search engine for Netscape. Does this make sense? Who are the people we should be talking to?

Landing Netscape as a customer would clearly be a coup, but to serve such a customer, Page and Brin needed more engineers. The company quickly grew to seven people—Google Inc. was threatening to overrun Wojcicki's living space. "They were there at all times of the day and night," she recalled, and oftentimes their cars blocked her driveway. Nevertheless, "they were very considerate tenants." Wojcicki recalls the boys helping Silverstein push his old Porsche 911 down the driveway and into the street at three in the morning. The car was prone to loud backfires upon starting, and the team didn't want to wake her.

But Google inevitably outgrew its first office space. In the spring of 1999, the company took up residence on University Avenue in the heart of Palo Alto. With a real lease and nearly ten employees, the new business needed a model for generating cash, and that meant it needed a salesman. Shriram recruited Omid Kordestani, a talented executive he knew from his Netscape days. After running through a gauntlet of four-hour interviews with Page and Brin—Kordestani recalls being grilled in what he called an "almost academic fashion"—he joined in early March as the first true business hire. Of course, it helped that before he earned his MBA (from Stanford, of course), Kordestani had earned an undergraduate degree in electrical engineering. With Shriram and Kordestani's aid, Page and Brin began plotting their strategy for bringing real money—and real visibility—into their young company.

The Biology Major and the VCs

In March 1999, Salar Kamangar was finishing his second degree at Stanford, in economics. He had already completed his first, in biological sciences, but had decided he didn't want to be a doctor. And who could blame him? All anyone at school was talking about was the Internet start-ups that originated on campus—Jerry Yang and David Filo had done it with Yahoo; Joe Kraus and his buddies had done it with Excite. Kamangar was eager to join one.

It seemed everyone had a start-up idea, including Kamangar (his had to do with online advertising), but he was smart enough to know that he needed experience first. So he headed over to a start-up fair on White Plaza, the center of campus activity at Stanford. Kamangar had been using the Google service for a while, and he had heard that the founders would be there. Like most early users, Kamangar thought Google provided much better results than either Yahoo or Excite. Could lightning strike a third time?

Sergey Brin was manning the Google booth that day, and Kamangar impressed him. "They only had engineering positions open," Kamangar recalled, "but Sergey promised to watch out for my résumé if something else opened up." Kamangar persisted and managed to land an interview at Google's University Avenue offices. He offered to work for free—he just wanted the experience. Brin was sold, and took Kamangar in as employee number nine—though he insisted on paying Kamangar an hourly wage.

It turned out Brin had a project for Kamangar: Ram Shriram had lined up meetings with a slew of Silicon Valley venture capitalists, and Google needed to put together a presentation which would impress the notoriously demanding financiers. Brin assigned Kamangar his first task: pull together the presentation. The biology major had two weeks to make it happen. "I was shocked and excited to be in the middle of it all," said Kamangar, now director of product management for Google.

Kamangar worked with Page and Brin to bang out a presentation based on a live demo. At this point in its young life, Google did not have a fleshed-out business model, but the prevailing method of making money from search at comparable companies like Yahoo was sponsorship and banner ads. Given Google's already impressive page views and prodigious growth (Kamangar estimated that the site was growing at nearly 50 percent a month), it was not hard to make a case that were Google to take banner advertising, it would be instantly profitable. Coupled with Google's clearly superior technology and star-studded lineup of angel investors, the presentation was a hit.

As this was early 1999, the Internet bubble was in full swing. Venture funds were swollen with money, and despite the fact that Google had no intention of becoming a portal, any deal with an Internet profile was in high demand. Page and Brin had a number of investors to choose from, and the firms they selected cemented Google's image as a unique company in the Valley. Page and Brin persuaded two of the most competitive top-tier firms—Sequoia Capital and Kleiner Perkins Caulfield & Byers (KPCB)—to take the deal together. KPCB had already invested in AOL and Excite, while Sequoia was already an investor in Yahoo. The firms led a $25 million round at a valuation of $100 million (several smaller players also participated in the round). KPCB partner John Doerr—famous for funding Amazon, among many others—and Sequoia partner Michael Moritz, who funded Yahoo, both took seats on the board.

When two of the most visible financiers in the Valley take a deal together, everyone in the industry takes note. The $25 million round marked Google's arrival in the Valley. "When this deal happened, it launched Google into a class of its own," said Ron Conway, an angel investor in the deal.

Michael Moritz, however, recalls his reasons for investing as more calculating. "The investment was done in part to help Yahoo," he recalls. "It certainly wasn't because there was a business model. At that time Yahoo thought of search as something [it] could outsource. When we looked at Google, the idea was that it would power a lot of other sites, most notably Yahoo."

Regardless of the initial reasons Sequoia or Kleiner invested, Brin and Page now had a $25 million war chest. To celebrate, they revisited Burger King and had a meal together, just as they had when Bechtolsheim invested.

Google was now on the map, but the company's extraordinary run had barely begun. Around this time, Terry Winograd received an e-mail from a Stanford administrator, asking about Larry Page's office space. All graduate students in the computer science department were assigned office space, and while Page and Brin were officially on leave, they still kept their connections to their alma mater via their offices. The administrator was wondering whether Page and Brin would be back for the fall semester. Winograd forwarded the e-mail to Page with the question "Are you coming back in the fall?"

Page's response: "I think it is kind of unlikely that I'll be back that soon."

"I remember the day they cleaned out their offices," Winograd recalls, adding that it took Page and Brin another year to actually leave Stanford. "I remember that day because they were very disappointed. They had this grim look on their face[s] because they had to go to Stanford with empty boxes, and leave with them full."

New Roles, Little Revenue

With the funding finalized in June 1999, Brin and Page found themselves in new roles: leaders of a start-up expected to bring significant return to its investors. Venture capitalists are well known for ruthlessness when it comes to protecting their money. As insurance, they often install their own people in the CEO position, pushing aside the founders in the process. Doerr and Moritz insisted that the company quickly identify and recruit a new CEO to replace Page, much as Tim Koogle had replaced Jerry Yang at Yahoo, or George Bell had replaced Joe Kraus at Excite. But finding a person that everyone could agree upon would not be easy. Page and Brin chafed at the idea of being told what to do by their new board members.

Regardless of the outcome of the CEO search, the new investors expected the founders to deliver a profitable business model. While they were at Stanford, Page and Brin had spent nearly all of their time improving the service. Increasingly, however, the founders were pulled into debates about business models, sponsorship deals, partnerships, and even the prospect of going public—a preordained event for companies that took money from high-profile VCs during the late-1990s Internet boom.[12]

Despite Kamangar's advertising presentation to the venture investors, Brin and Page were deeply suspicious of blending advertising and search. Indeed, in their academic paper introducing Google, they wrote:

In our prototype search engine one of the top results for [the search term] "cellular phone" is "The Effect of Cellular Phone Use Upon Driver Attention," a study which explains in great detail the distractions and risk associated with conversing on a cell phone while driving. This search result came up first because of its high importance as judged by the PageRank algorithm, an approximation of citation importance on the Web [Page, 98]. It is clear that a search engine which was taking money for showing cellular phone ads would have difficulty justifying the page that our system returned to its paying advertisers. For this type of reason and historical experience with other media [Bagdikian, 83], we expect that advertising funded search engines will be inherently biased towards the advertisers and away from the needs of the consumers.

Over time, the founders have clearly made peace with their reservations about advertising, but back in the early days, they were adamant that their company not fall into the same trap as had the companies that spurned them. Google would never put advertisers ahead of its users.

"We were motivated to have the best possible search no matter what," Brin recalls. "At the time that meant that if you had a banner ad, which was by far the easiest way to generate money off of search, that would mean that the load and render time of the page would increase significantly. We were interested in avoiding that. We also felt like, well, the ad has nothing to do with the search. Why would we show it? It's distracting."

This allergy to advertising, as Moritz phrases it, left the company searching for a sustainable business model. Given that the founders had slammed the door on portaldom—pretty much the entire business model of the consumer Web—the company was forced to try different approaches to making money.

The founders settled on an enterprise or original equipment manufacturer (OEM) model—Google would become a provider to the larger sites interested in furnishing superior search results. Kordestani was tasked with cutting deals across a broad swath of early Internet players, but he found the going extremely tough. Deals were few and far between—an early win, Red Hat software, came in at a paltry $20,000. Kordestani did land Netscape as a partner, but the deal did not push the young company into the black.

Press coverage of Google often glosses over this fact, but the truth is that the company lacked a viable plan for making money until early 2001. "There was a genuine concern (at the board level) about where the revenues were going to come from," says Shriram.

"We really couldn't figure out the business model," adds Moritz. "There was a period where things were looking pretty bleak. We were burning cash, and the enterprise was rejecting us. The big licenses were very hard to negotiate.

"As 1999 trickled by and we were burning cash without a clearly illuminated path to revenues, there was considerable concern," Moritz continues. "The benefit Google had was that it had fairly low burn rate compared to the behemoths [like Yahoo]. We had enough cash, but it always rattles people when hundreds of thousands of dollars a month go up in smoke and there is no bread on the doorstep."

The story of how Google found its business model—and its subsequent rise to glory—requires a diversion into the history of another company, GoTo.com. For while Page and Brin struggled with the notion of turning search into a business, the founder of GoTo.com, Bill Gross, saw in search the seeds of an economic revolution.

Chapter 5

A Billion Dollars, One Nickel at a Time

The Internet Gets a New Business Model

Advertising ministers to the spiritual side of trade. It is great power that has been entrusted to your keeping which charges you with the high responsibility of inspiring and ennobling the commercial world. It is all part of the greater work of the regeneration and redemption of mankind.

—Calvin Coolidge, to the advertising industry

Had he just stuck to his guns, he'd be the one hailed as the revolutionary, the one on the cover of every business magazine, no, the cover of *Time* magazine, with a guest chair on *Charlie Rose* to boot: Bill Gross, founder of the company with the most anticipated IPO in the history of Wall Street, the mad genius who rewrote the rules of business and rewired the way our culture understood itself.

Indeed, had Bill Gross not given up his argument, had he just followed his gut, there might not even be a Google. Brin and Page might have sold out to Yahoo or Excite or Microsoft, or merged with Ask Jeeves, or gone the way of AltaVista—sinking slowly into the dark oceans of corporate M&A. Imagine that, *a world with no Google.* A world where Brin and Page, those arrogant little upstarts, are no more than forgotten footnotes in a much grander story—

the story of a serial entrepreneur with a mottled past who finally proved himself beyond all possible doubt. Indeed, had this version of history come to pass, this very book would be talking about how GoTo "transformed our culture."

Only it's not. Bill Gross has not created tens of billions of dollars in market value, at least not yet, and the trail of lawsuits and querulous press clippings littering his past are proof that he failed in his quest to get each and every one of his investors fuck-you rich. But Bill Gross can quite legitimately claim to have created the business model that made Google possible, in the process reinventing pretty much the entire economic cardiopulmonary system of the Internet. And at the end of the day, that's certainly something.

Wiry, manic, and bespectacled, Gross is philosophical about the matter. Brimming with a conspiracist's good-natured glee, he's eager to pull you into his confidence. After all, while most people have never heard of the man, the company Gross founded later became Overture, a paid search giant sold to Yahoo in 2003 for more than $1.6 billion. Not a $30 billion IPO, but not pocket change, either.

Parallel Entrepreneur

By his own account, Gross has been starting companies since he was thirteen. His problem was never ideas. No, he, in fact, has way too many of those. His problem was scale—how could he possibly start companies as quickly as he could dream them up?

Gross started in a linear fashion, building companies one at a time. He'd grow them till he got bored or distracted (or both); then he'd sell them. He funded his first year of college by selling solar energy conversion kits through ads in the back of *Popular Mechanics*. While still an undergraduate (at the California Institute of Technology in Pasadena), Gross hacked up a new high-fidelity speaker design and launched GNP, Inc., to sell his creations (GNP stood for

Gross National Products—an indication of Gross's sense of humor as well as an underdeveloped sense of modesty).

But Gross had reason to boast: GNP, Inc., grew to claim number seventy-five on *Inc.* magazine's 1985 list of the 500 Fastest-Growing Companies. When he graduated, he sold the speaker business to his college partners and started a software company that presaged much of the rest of his life's work. The company, GNP Development, allowed computer users to type natural language commands that the computer would translate into the arcane code needed to execute specific tasks. In other words, Gross's company created a program that in essence let you "talk" to the computer in plain English, as opposed to computer code. Gross's program was a small step toward Silverstein's *Star Trek* interface (as discussed in Chapter 1)—the holy grail of nearly everyone in search today.

Gross's program worked with just one application, Lotus 123, the precursor to spreadsheet titan Microsoft Excel. It turned on a tantalizing idea: imagine the day when you could talk to your computer in plain English, and it would understand and execute your commands! Gross's approach was, in essence, a neat hack, the kind of thing Ask Jeeves tried (and failed) to do in the search business a decade later. Because Lotus 123 was a limited environment with a structured set of input commands, Gross and his programmers could pretty much deduce most of the natural language that a user might come up with. (You weren't going to ask Lotus 123 for photos from the Mars Rover, after all.)

But GNP Development illustrated another side of Gross: he is a man willing to bend the rules of acceptable business behavior to see his visions become reality. When the folks at Lotus realized that GNP was onto something (about the time GNP hit a million or so in sales, according to a 1998 *BusinessWeek* report), Lotus sued. The reason: GNP's packaging was a bald copy of Lotus 123's look and feel, and Lotus didn't appreciate GNP's turning tricks while wearing Lotus's trade dress. But despite his faults, Gross is a hard man to hold a grudge against, and he managed to convince Lotus that GNP was good for the tech giant. Lotus not only dropped the lawsuit; it bought GNP for $10 million. Bill Gross had made his first fortune.

Lest his role as an innovator be obscured, it's worth restating this fact: in 1985, Gross was already working on a major piece of the search problem—a natural language interface. And after his company was sold to Lotus, Gross stayed on, because Lotus offered him the chance to focus on another aspect of the search problem: indexing.

Now, back in the 1980s, there was no Web to index, but there was the personal computer hard drive. And while PCs held a mere 20 or 40 megabytes of data at that time, most were already a mess of lost files and hopeless organizational structures. What the PC needed was a search engine, and that's why Gross invented Magellan.[1]

Magellan was an early version of what is now known as a file manager, a way to "search all your files on your hard disk instantly," Gross explains. Sounds simple, but in the mid-1980s, this was a pretty revolutionary idea. Magellan flattened out the file system, putting all files across DOS directories in one big view. It quickly garnered thousands of fans, but languished after Lotus shifted focus from spreadsheets to its Lotus Notes groupware application.

As Magellan withered, Gross grew bored with life at a large company. At the same time, he realized his young son was growing up. So in the early 1990s, he started a new company, Knowledge Adventure, which focused on software that helped kids to learn. Once again, Gross was working on a piece of the search problem: this time, how people learn (the more you know about that, the more you can program a machine to help people ask questions).

The company took off, becoming the world's third-largest children's software publisher. But Gross was not cut out to run a large company, as it provided no outlet for his voluminous ideas and endless energies—in fact, had he not left, many colleagues say he would have been booted out by the board. But Gross did leave, and in 1996 Knowledge Adventure was sold to Cendant for $100 million.

Bill Gross had made it to the big leagues, and his fortune had multiplied tenfold. But he was frustrated with the cycle of creating, building, then selling companies. Through Knowledge Adventure he had met and befriended director Steven Spielberg, and he was fascinated with the way Spielberg ran movie sets. "He walks around all day using his brainpower to creatively enhance things around him," Gross told *Inc.* magazine in 1997. "I'd always thought you had to take the good with the bad. How audacious to think that your job could be perfect all day long. But here was someone doing it."

Inspired by Spielberg, Gross decided his dream job was to start a company that allowed him to start many companies in parallel—a business incubator of sorts, an idea factory. The Internet was just starting to take off, and Gross had far more ideas than time to execute them—and all of them, he believed, could work. It was just a matter of time (never enough of it) and people (never enough good ones). What he needed was a company that compressed time and leveraged people, a company that let businesses be conceived, prototyped, and launched quickly. And so in 1996, IdeaLab was born.

The Idea Factory

Spend an afternoon with Bill Gross in the IdeaLab offices, and you'll get the sense that had he not created IdeaLab, he might have self-destructed. IdeaLab is his protective shell, his habitat, his carefully tended nest—it contains his ideas, gives structure to his bouts of creative energy, allows him to breathe.

IdeaLab was (and remains) a business incubator, but given its birth at the onset of the Internet boom, it quickly became far more than that. For a brief moment, IdeaLab was a major hub not only of the Internet industry, but of cutting-edge business theory to boot. Gross theorized that the true value in enterprises lay in people, and that the laborious process of starting businesses—from hiring to finding office space—didn't allow capital to efficiently realize that value. At IdeaLab, great people would be given the space, resources, and support needed to realize their ideas, and if an idea failed, that was OK; the team would move on to the next one. No muss, no fuss.

"In my earlier businesses I was always looking to assemble the right team," Gross explains. "I thought, *Wouldn't it be great if you didn't have to do that every time you had a business idea?*"

Gross set out to build teams that could incubate businesses quickly. IdeaLab began rapidly prototyping his profuse outpouring of ideas and—in theory anyway—pushed only those businesses that could succeed out the door and on to greater glory as public companies. IdeaLab seeded each company to a maximum of $250,000, made introductions to other VCs, then kept a minority interest. As Gross was fond of theorizing at the time, one big hit would fund IdeaLab forever.

Early on, it certainly seemed as if Gross would have his one big hit, and then some. A partial listing of the companies IdeaLab created reads like a to-do list for the Internet economy, circa 1998: FreePC (giving away PCs on the idea that Internet services would pay the bill on the back end), CitySearch (local listings and information), Tickets.com (selling tickets over the Internet), and eToys (the Amazon of toys), among many others. Gross even launched answers.com—a search engine "powered by humans." Sound familiar? Yep—it was Google's Google Answers service, circa 1998.

The investing world loved Gross's ideas, and for a while anyway, it loved his companies as well. Ben Rosen, the former chairman of Compaq, was an investor in IdeaLab and told *Inc.*: "There are very few examples of entrepreneurs who have started more than one successful company—it's really hard to think of any that have had two big hits. Bill has a chance of having a dozen hits. I think in five years' time Bill Gross will be as much of a household name as any household name in technology, even though today he's barely known outside of a very small circle."

Five years later, of course, Google was the household name. But in 1998 and 1999, many of IdeaLab's companies went public in spectacular fashion, and on paper, Gross and his investors got very, very rich. IdeaLab was widely imitated as a model, as were its companies (IdeaLab had one of the first online pet supply companies, for example, as well as the first online cooking site). In a very short time, IdeaLab took in more than a billion dollars in capital from an impressive slate of high-profile funds and individuals, built dozens of businesses, and had filed plans for its own IPO valuing itself at an astonishing $10 billion. But like so many leaders of the early Internet era, Bill Gross was smoking a little too much of his own stuff, and the party came to an abrupt and unhappy end.

"For a while there it seemed like we could do an idea a month," a somewhat chastened Gross tells me. "As long as the updraft was continuing, it worked." But the updraft ended, the capital markets stopped funding concept plays, and by the middle of 2001, IdeaLab investors were left holding a shattered portfolio. They eventually filed suit, demanding that Gross liquidate IdeaLab and all its holdings, so they could at least get some of their money back. For they saw in the wreckage of IdeaLab one shining gem that could help them recoup at least some of their losses, one company that was growing like a weed despite the carnage of the dot-com bust: Overture.

GoTo.com: A New Model for the Web

If Google is a grand slam, then Overture was a triple ripped through the gap: good, but the base runner didn't quite get home. Founded in late 1997 as GoTo.com, Overture remains Bill Gross's greatest financial success—a company he built and sold not for $10 million, or even $100 million, but for well over a billion dollars. Given the scale and scope of such an achievement, you might expect Gross to be ecstatic when discussing his prodigy. Instead, a tone of regret and a tinge of pain shade his recollections, evidenced by small hesitations in his otherwise exuberant demeanor. Overture was a hit, yes, but it might have been Google, or at least it could have tried to be.

At the core of Gross's insight was the premise that search was broken, but the portals didn't seem to care. Google later proved that search mattered, but when GoTo launched, Google was still an obscure graduate school project, and conventional wisdom said search had already had its day. By the time GoTo debuted, the market was in full-blown portal madness. Search was "good enough," Louis Monier told me in 2003, recalling the declining days of his brainchild AltaVista with more than a hint of disdain in his voice.

Search became a problem of sorts: executives knew that when someone searched the Web, chances were he'd leave the portal if he found something that matched his intent. Hence, it wasn't in the portals' interest to improve search results. Sites that had built their audience and traffic on search—AltaVista, Yahoo, Excite, Netscape—shifted strategy and began to act like media properties jealous of their audience. (In fact, Tim Koogle, CEO of Yahoo at the time, went so far as to brag in an analyst meeting that his search-related traffic was *declining.*)

To further consolidate their traffic dominance, the portals parlayed their overheated stock currency into an acquisitions binge, buying anything that promised to extend their ability to be sticky—e-mail services, video services, home-page building services. By the late 1990s, the entire Internet world was in play. Yahoo, for example, purchased Geocities, Broadcast.com, Four11, ViaWeb, and several others, for a total of nearly $10 billion between 1998 and 2000.

As the portals consolidated their grip on Internet traffic, demand for that traffic from independent e-commerce players soared. Acquiring traffic became expensive—the major portals charged millions of dollars for real estate on their sites, and Internet companies, flush with VC and public cash, lined up for the right to be there. The litany of traffic deals in 1998 and 1999 reads like a dot-com death march: CDNow spent $18.5 million for a deal with Lycos; Preview Travel $15 million for real estate on Excite; AutoConnect $17 million with AOL.

The premise for this slew of traffic deals was thin: the e-commerce sites were buying access to customers without much sense of whether those customers had any interest whatsoever in what the sites were offering. While it might seem sexy to be the exclusive credit-card marketer for AOL, or the premier CD purveyor to Lycos, proof that such deals amounted to much of anything was hard to find. In fact, warning signs were everywhere—the normally exuberant research outfit Jupiter Communications published a report in 1999 detailing the failure of such deals. But this was a land grab, and in such an environment no one was looking to be left out. The ethos of the age: get in on the ground floor first, outspend the competition, and hold on for dear life.

For a brief moment (and it was brief, in the scheme of things—less than three years), accountability mattered little. The entire Internet space became a Super Bowl buy. What CDNow did with the traffic it got was less important than the fact it had the traffic in the first place. To many in the industry, traffic was a universal lubricant justifying Internet valuations. In the late 1990s, several companies received venture funding and/or managed to go public simply by acquiring rights to real estate on portal sites like Netscape or Yahoo.[2]

As a result, innovation in search languished, and the tragedy of the commons prevailed: spammers quickly took control of the indexes. Search-engine spam—irrelevant listings pushed up the index by bad actors looking to acquire free traffic—remains a major problem to this day. But although today's major engines are increasingly sophisticated in their approaches to combating spam, in 1998 search-engine spam was barely even understood.

Before Google, most engines employed simple keyword-based algorithms to determine ranking. While the actual computer science is a bit more complicated, in essence they indexed the words on a particular page, then matched those words to search phrases. It worked great for small, controlled data sets, and as AltaVista proved (see Chapter 3), it worked quite well for the early Internet. But once spammers (the adult-entertainment industry in particular) realized they could capture traffic for high-traffic keywords like "cars" by hiding those keywords all over their sites (often in small white letters on a white background, for example), the model quickly broke down. This is why, by late 1998, the majority of results matching a search for "cars" on Lycos were porn sites.

Gross Sees an Opening

Bill Gross watched spam gum up listings on the major engines, and he surmised that the only way to combat it was to attach some kind of inherent value to the process of searching. "Search makes markets more efficient," Gross tells me. "But by 1998, the spam in search was so extreme it wasn't working anymore."

Without an economic price associated with listings, he reasoned, spam would overrun the system. Force the friction of pricing into the equation, and the markets would start to behave rationally.[3]

As spam worked its tendrils through the lattices of nearly every major search engine, executives at the major portals simply ignored it, as did the mainstream press, save the odd rejoinder about porn. In effect, the market had stopped valuing the very mechanism that was proven to drive traffic in the first place. As stickiness became all-important and as raw traffic metrics became the new currency of the Internet boom, an opportunity opened up. Gross knew that the e-commerce sites buying advertising on the portals were failing to justify their expenditures. And he thought he knew why.

Gross sensed there was a massive difference between *good* traffic—traffic that converted into paying customers or loyal users of a service—and *undifferentiated* traffic: people who had come to a site because of spam, bad portal real estate deals, or poor search-engine results. At the time he was developing GoTo, Gross had more than a dozen other Internet-related IdeaLab companies in various stages of execution, and all of them needed good traffic—customers who were, in fact, interested in the products or services his companies were offering. *How,* he wondered, *can one differentiate between good traffic and crap?*

Gross became obsessed with garnering qualified traffic for his businesses, and he developed GoTo.com with an eye toward solving that problem: none of his companies could afford multimillion-dollar deals with portals like AOL or Yahoo, and in any case Gross sensed, correctly, that those deals would probably yield more bad traffic than good. How might an online business like CarsDirect or CitySearch buy the traffic it needed, when it needed it, at a cost that made sense for that business?

Solving this problem became GoTo's mission. Gross studied his IdeaLab companies' traffic acquisition numbers and computed the costs of each company's campaigns down to the single visitor. He noticed that with proper maintenance, IdeaLab could buy decent traffic for its sites from various ad networks, running traditional banners, for between seven and ten cents a click, or visit. When he got really good at managing his campaigns, he could drive that price per click to five cents or even less. In other words, Gross noticed that traffic could be had for pennies, if you worked hard enough at it.

"We used this great software to monitor all our traffic acquisition efforts," Gross recalls, referring to Flycast, an advertising network and cost-per-click tracking service that, like so many now-defunct Internet companies, was about five years ahead of its time. As Gross watched the metrics dance before his eyes, he began to sense what might be called a true price each of his companies would be willing to pay to obtain the right kind of visitor—and he realized that his true price was far higher than the cost of obtaining traffic through conventional banner advertising approaches.

Put simply, it's not the quantity of traffic, Gross realized; it's the quality. Any business would be willing to pay a lot more than seven to ten cents a click for the *right* traffic!

That realization became Gross's eureka moment—a moment

that, more than any other, spawned today's Internet advertising economy. For every single online business (even, it turns out, portals), undifferentiated traffic is worth very little, but specific traffic, traffic with *an intent to act in relation to a business's goods or services,* is worth quite a lot. Gross realized that businesses will pay quite a bit to acquire the right kind of traffic. All he had to do was build an engine that created intentional traffic. And here's where it all fit together: the Internet already had a model for an engine that created intentional traffic. It was called a search engine. Only nobody seemed to care about it anymore!

Energized by his insights, Gross set out to build a better search engine, one that would both defeat spam and produce insanely relevant results. Together with his IdeaLab team, Gross looked at human-edited approaches, as Yahoo had done early in the Web's history, but found they couldn't scale to Internet proportions. He tried finding better algorithms (the approach Page and Brin were tackling four hundred miles to the north at Stanford), but Gross was convinced that any approach to search driven by algorithms would ultimately be outsmarted by spammers (to this day, whether that assertion is true remains an unanswered question). No matter what approach Gross tried, he felt the endgame was no better than the spam-choked, irrelevant engines of the day.

So Gross turned to his original idea: to kill spam, one must add the friction of money to the equation. But how? Certainly you can't charge the Internet user for searching. But what if you could charge the advertiser?

Gross's core insight, the one that now drives the entire search economy, is that the search term, as typed into a search box by an Internet user, is inherently valuable—it can be *priced.* "All our false starts made me realize the true value of search lies in the search term," Gross says. "I realized that when someone types 'Princess Diana' into a search engine, they want, in effect, to go into a Princess Diana store—where all the possible information and goods about Princess Diana are laid out for them to see." GoTo.com was to become a mechanism for those stores to get built, one keyword at a time.

At least, that was the theory. But to get all those merchants to participate in the grand GoTo experiment, Gross would have to somehow persuade them to give the new engine a try. And that's where a brief detour into the economics of candy arbitrage becomes necessary.

The Sugar Daddy: It's All About Arbitrage

When he was twelve, Gross lived in an apartment building in Encino, California, outside of Los Angeles. There were hundreds of kids in that complex, Gross recalls. "We all roller-skated together, played baseball together, swam together, did everything together," he tells me. And when they had saved up enough money, they all made the pilgrimage to a local pharmacy, where they'd buy their fix of candy. "We used to hop the cinder-block wall surrounding the complex and go buy candy for a dime at the West Valley Medical Center," he recalls. "We'd go there all the time."

Now here's where it gets interesting. In Gross's words: "One day I was at Savon [pronounced Save-on] on Ventura Boulevard and saw they had a special on candy, three for a quarter. So I bought five dollars worth—at eight and a third cents each—and brought them back to my apartment, where I sold them for nine cents. I saved the kids a penny, *and* they didn't have to hop the wall. *Everyone* began buying from me. I would ride my bike up there to get the candy and bring it back in bulk in a big Styrofoam cooler box I mounted on the back."

In essence, Gross staked an initial capital investment of five bucks on an arbitrage opportunity in the local candy market, and it paid off. He was making two-thirds of a penny on every unit—roughly an 8 percent margin—but he really started cleaning up as his volume increased. "After I started buying whole boxes of candy, Savon sold it to me for seven cents. And then finally, when my volume got really big, and I was selling at the bus stop and at school in the mornings, I got it for six-point-four cents, as I recall, from Smart and Final in Van Nuys."

Volume had driven Gross's margin up from 8 percent to more than 40 percent. With the profits, Gross paid for his next project: the solar energy kits he sold in the back of *Popular Mechanics*. "I made a business in candy that allowed me to buy the math books and solar energy parts I wanted," Gross explains. Those kits, in turn, paid Gross's way into Caltech.

Gross learned several things from his days as a player in the candy trading market: first and foremost, it pays to be a supply-side sugar daddy in the middle of a high-demand transaction with clear market imbalances. Second, Gross realized that you can make significant money on pennies a transaction, if the volume is high enough. And third, he developed a taste for entrepreneurship, a taste he has clearly never lost.

What Gross spotted in the frothy search market of 1997–1998 was another arbitrage opportunity. As defined in *Webster's,* arbitrage is "the nearly simultaneous purchase and sale of securities . . . in different markets in order to profit from price discrepancies." Gross observed that the market for any kind of traffic—be it undifferentiated or intentional—valued clicks at about five to ten cents each, but it seemed obvious that the inherent value of intentional traffic should be far greater. If Gross could harness and sell a search engine's ability to turn undifferentiated traffic into intentional traffic, he'd make a killing on the spread.

But Gross had a conundrum. To launch a search site like GoTo.com, he needed both audience and advertisers—and the more advertisers the better. (GoTo filled out its search offerings with a standard organic search feed from Inktomi.) Gross knew he could buy his audience, and he reasoned he could arbitrage that audience's intentional traffic—as reflected in the keywords they typed into his engines—against an advertiser's desire for business. But he needed a critical mass of keyword-buying advertisers to support his site, and given the untested and relatively complex nature of what Gross was creating, it was going to be quite difficult to persuade those advertisers to come on board and bid for keywords. After all, while Bill Gross understood the intrinsic value of a keyword, not many others in the Internet world did. Until he could prove otherwise, Gross was selling theory, and little else.

Gross solved his problem by adopting the time-honored approach of dumping—or perhaps drug dealing is a better comparison: the first one's free (or nearly so). Gross built not one but two entirely audacious ideas into GoTo's initial business proposition for advertisers: first was the concept of a performance-based model—one in which advertisers paid for a visitor only when a visitor clicked through an ad and onto the advertisers' sites. Instead of demanding upfront money from advertisers, the way AOL or Yahoo did, GoTo.com's model guaranteed that advertisers had to pay only when their ads were clicked upon. Of course, this is now the standard model for the multibillion-dollar paid search market.

Second, and even more audacious, was how Gross priced his new engine: one cent per click, an extraordinary discount to the market. He knew his price was seven to ten times less than what every Internet marketer was paying at the time, and in an environment where traffic was crack, advertisers couldn't help but look to Gross for a fix.

In short, Bill Gross bought traffic from one place for five to ten cents, and resold it on his site for a penny. Not exactly a great business model. But Gross believed that the market would take over, and that soon advertisers would compete to be listed first for high-value keywords like "computer," "camera," and book titles. On the come, Gross was betting that market forces and the greater value of intentional traffic would push per-click prices past his cost of traffic acquisition.

Gross's gamble lay in building out GoTo as a habit for both his advertisers and his audience. Back at the IdeaLab's headquarters, he built out elaborate models showing how GoTo would slowly grow audience and advertiser share, and how his plan of arbitraging traffic would eventually turn profitable as advertisers began to bid various keywords up from one cent to as high as two dollars.

"Eventually, with volume, I was able to drive traffic acquisition costs down to six and sometimes four cents," Gross recalls. "Then people would exit paying a penny, or possibly two, as some might click on more than one link," he continued, warming to his tale. "But people were also bookmarking the site, and using it again, which drove down my average cost to acquire a searcher/search. With volume and loyalty, my cost to drive a search was declining each month, and my earnings for each search were increasing."

In about six months, Gross claims, the two prices met and crossed—the average price paid by an advertiser rose past the average price GoTo paid to acquire a searcher. "Our model had them crossing in about two years," Gross says, "so we were way ahead of schedule. I was certain we could get there, because I knew bid prices would increase to their true value over time, and I knew the true value was somewhere in the [range of] twenty-five cents per click to two dollars fifty cents per click and even higher on some terms. I never knew some would go to one hundred dollars [as they have for terms like "mesothelioma," a rare cancer that—in a gruesome twist of capitalist fate—affords a high chance of recovering damages in a lawsuit], but I was sure they would beat one dollar or two dollars, and they did."

Back in 1998, the idea of basing a business on the idea of pay per click was viewed as a wild and rather dismissable gamble. After all, if you're Yahoo or AOL, why would you ever want to be held accountable for the performance of what you sold to your partners? If marketers couldn't turn the traffic into profits, that was someone else's problem.

"The more I [thought about it], the more I realized that the true value of the Internet was in its accountability," Gross tells me. "Performance guarantees had to be the model for paying for media."

Gross knew offering virtually risk-free clicks in an overheated and ravenous market ensured GoTo would take off. And while it would be easy to claim that GoTo worked because of the Internet bubble's ouroboros-like hunger for traffic, the company managed to outlast the bust for one simple reason: it worked. For consumers, GoTo provided relevant, if commercial, results, but most users went to GoTo for commercial results in the first place. For advertisers, GoTo's model was a dream; for pennies a click, they could bring traffic to their site, and oddly enough, the traffic that came seemed to be the best kind: actual customers who stuck around and either purchased products or became regular visitors to their site. Hell, reasoned marketers, if each click brings paying customers, I'll pay as much as I can afford to bring 'em in.

An Inauspicious Launch

In February 1998, Gross introduced GoTo.com at the famed TED (Technology, Entertainment, Design) conference in Monterey, California, before an elite gathering of seven hundred or so high-tech influencers.[4] Gross was in high-visionary mode for his presentation ("He always gave great demo," commented Lotus founder Mitch Kapor), but once Gross got started the crowd of usually enthusiastic boosters became confused: Gross was pitching a new search engine (they understood that), but the results were driven not by an impartial crawl of the Web (as AltaVista was), but rather by whoever paid the most to be associated with the searcher's keyword or phrase.

In short, the cognoscenti at the TED conference did not approve. The hallway chatter after Gross's presentation painted GoTo as intellectually interesting but a bit loony. Not only was a pay-for-placement search engine seen as technically problematic; it was in clear violation of every ethical boundary known to media. GoTo was putting the advertising peanut butter into the editorial chocolate, and the press largely echoed the cognoscenti's review, framing the debate as one of editorial purity: a search engine where the results were bought and paid for—imagine if *our* periodical had such practices![5]

Gross defended his brainchild vigorously, noting that in the GoTo model, the marketplace was transparent: consumers were actively informed of which advertisers were paying for what keyword, and even how much (on initial versions of its site, GoTo listed how much advertisers were willing to pay for each click). As the press storm continued through 1998, Gross stuck to his guns, arguing that GoTo was akin to a yellow pages for the Internet: those who paid for larger ads got more calls. And just as they did for the yellow pages, visitors who came to GoTo came with an intent to buy. GoTo.com was a commercial search engine, an engine of purchasing intent.

The yellow pages metaphor stuck, and it became something of a mixed blessing for GoTo—on the one hand, it got advertisers and customers comfortable with the new search engine (what could be more innocuous than the yellow pages, after all?). But on the other hand, the metaphor ignored the more subtle and complex market truths Gross believed lay at the heart of GoTo's proposition. For GoTo was not just the yellow pages; it was the yellow pages crossed with the NASDAQ stock exchange. Pricing wasn't fixed; it was determined by an ever-present, transparent, and accountable market valuation process. Gross was one of the first to see a world where millions upon millions of search queries created the perfect advertising marketplace, and like a missionary, he preached the GoTo gospel to whoever had the patience to listen.

When the GoTo.com service launched (four months after TED in June 1998), it sported just fifteen advertisers. But within six months it had hundreds, and by 1999 its advertisers numbered in the thousands. Gross had created a platform that let his advertisers build his business. This was a revolution indeed: a timeworn maxim of the advertising business, attributed to John Wanamaker, a department store owner, declares that you know you're wasting half your advertising budget; you just don't know which half. With GoTo, there was no waste.

The Rise of Syndication

GoTo may not have been a darling of the press, but it prospered nevertheless. Within six months of launch, GoTo had taken root. Gross and his team—he had hired a colleague, Jeffrey Brewer, as CEO—knew he was onto something. The company's network of advertisers grew to nearly eight thousand by the middle of 1999, and revenues were on pace to surpass $10 million annually by 2000. While the company was not yet profitable, Gross's arbitrage bet was beginning to pan out. GoTo.com was serving more than 100 million searches a month, with about 10 percent of those resulting in clickthroughs, or what GoTo labeled "paid introductions."

On the strength of metrics like these, GoTo filed to go public in April 1999. As its filing shows, GoTo executives had begun to feed their arbitrage engine through traffic acquisition deals with major sites—in essence, spinning undifferentiated straw into pay-per-click gold. In the course of its first full year of operation, for example, GoTo purchased 180 million clicks from Microsoft for a total of $6 million—or about 5.5 cents per click. It also negotiated a series of deals with Netscape to provide traffic to GoTo at an average of about four cents a click. At the same time, GoTo executives realized they could extend their network by syndicating GoTo's PPC model to a host of other sites across the Web. In short, GoTo would provide search services on other companies' Web sites in exchange for a fee or a split in revenues.

As a result, GoTo developed two lines of business: its main site, GoTo.com; and a syndication business, which had lower margins (Gross had to split the revenues with his publishing partners) but far more scale. Gross's decision to syndicate his listings was a critical one—by offering his service to other search engines, he picked up important new distribution channels, which in turn extended the reach of his advertising network. That in turn increased the number of advertisers who signed up to use his service. GoTo prospered, and

Google executives took notice. Over the course of the next two years, they began to develop a response.

But in the middle of 1999, at a time when Google had arguably no business model to speak of, Gross had already positioned GoTo as the company to beat in paid search. His company executed a successful stock offering and continued to siphon undifferentiated traffic from major sites. Before too long, however, the portals began to take notice.

When they realized GoTo had essentially leveraged their traffic into a successful business, they decided they wanted a piece of the action. And that's when Gross and company cut what may well be the most important deal of their company's short history, with AOL.

Signed at the tail end of the Internet bubble in September 2000, the AOL deal was GoTo's largest and most significant syndication win. Its terms were reasonably simple: GoTo would pay AOL a whopping $50 million to syndicate GoTo's search listings on AOL's site. GoTo would make its profit on the traffic AOL sent through the GoTo listings. And profit it did. "The AOL deal was huge for us," says Ted Meisel, a McKinsey consulting veteran who took over as CEO of GoTo in May 1999. "As a company we turned a profit shortly after that deal began."

A Decision to Rue

The AOL deal triggered a round of soul-searching at GoTo. The company began its life as a destination site—Gross's original vision was of a massively scaled search site, AltaVista without the spam or irritating portalitis. But GoTo's syndication business was proving more successful, and it seemed to offer limitless growth. By the fall of 2000, GoTo's syndication network provided 90 percent more traffic than its destination site. GoTo.com, on the other hand, grew more slowly, and it faced significant competition from the very companies GoTo was now in partnership with on the syndication side.

But Gross argued that GoTo could do both. "We had heated debates," Gross remembers. "It was unclear which way to go. I thought we could get away with keeping focus on the site."

But GoTo's executive team worried that the company's syndication partners—AOL in particular—would balk at having to compete with GoTo's own destination site. And the concept of search as a portal unto itself was still unproven—Google had not yet broken out. With the dot-com bust deepening, GoTo's executives convinced Gross that the best course was to phase out GoTo's destination site in favor of the syndication business.

In September 2001, GoTo.com formally changed its name to Overture. The name change was reflective of what the company viewed as its core mission: making paid introductions (overtures) between visitors to its client Web sites and the company's vast network of advertisers.

But all along Gross was worried they were making a mistake. "We were worried about channel conflict and we overreacted," Gross says ruefully. "We thought that if we didn't phase out the GoTo.com site, our partners wouldn't renew. But the truth was, as long as we were making them money, they didn't care. We could have gone the destination route."

Indeed, just three months later, after GoTo had announced its new focus and its intent to change its name, Gross realized that Google was gaining ground—and as a pure search destination.

"They had just crossed over ten percent of all searches," Gross recalls, referring to the total percentage of Internet searches performed by the young service. "But they were not profitable."

So at yet another TED conference—this one in 2001—Gross met with Larry Page and Sergey Brin to suggest the two companies merge into a partnership that would once again realize Gross's dream of creating the ultimate search destination. But Page and Brin turned a cold shoulder to Gross's overture. The reason given: Google would never be associated with a company that mixed paid advertising with organic results.

The ghosts of Overture's past—and of the cognoscenti's snubs of TED in 1998—still clung to Gross and to his company. "They were so pure about advertising," Gross recalls of the Google founders, who by 2001 were royalty on the floor at TED. "We talked and talked, but nothing came of it."

Several months after the talks stalled, Google introduced AdWords, its answer to Overture. At first AdWords lacked a NASDAQ-like pricing element, but when Google adopted a pay-per-click model in early 2002, Overture sued for patent infringement (the case was settled right before Google's landmark IPO). But by then the horse was out of the barn. The new business model for the Internet had formally come of age—and Google, for the most part, was getting credit for it. To this day, Overture employees bristle at the mention of AdWords.

Adding injury to insult, AOL did not renew its $50 million deal with Overture, choosing to go with Google—even though, as a search destination, Google clearly competed with AOL for traffic. Gross had once again been proven right, and once again it was too late to do anything about it. In press reports, newly minted Google CEO Eric Schmidt called the AOL deal his company's "defining deal for paid listings."

Gross and other Overture executives claim Google actually lost money on the AOL deal so as to steal the business from Overture, but that claim is relative: as was typical for search deals in those days, Google partially paid AOL in pre-IPO equity, shares that as of this writing are worth more than a billion dollars. Not such a bad deal, after all.

With its unparalleled brand and traffic strength, and AOL in its back pocket, Google was now a force to be reckoned with. Overture countered by signing a paid listings deal with Yahoo and strengthening its deal with Microsoft, but the Wall Street analysts gave the economic edge to companies that controlled their own destiny—in short, companies that were in one way or another destination sites, just as GoTo had been. While Overture had impressive revenue and earnings growth—in 2002, the company earned more than $78 million on $668 million in revenue—Yahoo, Google, and MSN controlled the traffic flow. Overture was viewed, and valued, like a parts supplier: like Delco, to Yahoo's or Microsoft's GM.

The Search Economy Gathers Strength

As 2003 dawned and the IT business began to shake off the snows of a two-year winter, portals were once again king, but this time they understood the difference between good traffic and bad. Overture, though wildly profitable and responsible for defining and proving a business model that venture capitalist Bill Gurley praised as "the salvation of the Internet," was eclipsed by Google and overshadowed by its own partners Yahoo and Microsoft.

As the company surveyed its strategic options, its major competitors and partners were busy doing the same. Both MSN and Yahoo realized they needed to rethink their search strategies. To profit from search and control its own destiny, a company requires three elements, all of which Google already owned. First, it must have high-quality organic search results, also known as algorithmic, or editorial, search. Both MSN and Yahoo had outsourced these results to Inktomi or Google. Second, the success of Overture and Google's AdWords proved that a company needs a paid search network. Both MSN and Yahoo were outsourcing this element to Overture. And third, it needs to own its own traffic—the consumer's search queries against which editorial and paid results can be displayed. What Microsoft and Yahoo realized as 2002 came to a close was that this was the only element that either of them truly owned.

Overture also owned only one of these three magic elements—the paid search network. It lacked its own organic search technology, just as Microsoft and Yahoo did, and most important, it lacked a truly scaled destination site. Such sites were hard to come by, and even harder to build from scratch. Yahoo quickly moved to secure its own organic search technology, scooping up Inktomi in December 2002. Microsoft eyed both AlltheWeb, a European search company with impressive technology, and Ask Jeeves, a growing second-tier player.

But the giant could afford to wait and see, and it favored building its own technology, should it feel that the market had gotten big enough to justify the investment. In the summer of 2003, Microsoft decided to do just that, embarking on a massive internal search project, code-named "Underdog," to counter Google's growing dominance.

As for the paid listings piece of the puzzle, the writing was on the wall. Both Yahoo and Microsoft began to pencil out strategies for acquiring Overture.

Overture was in a pickle. If either of its major customers decided to bail, it would lose a significant amount of market share and its stock would tank. If it moved to purchase or build a portal, its partners might balk or, worse, bolt to Google, as AOL had. This did not put Meisel in an easy position from which to negotiate a deal. Both Yahoo's Terry Semel and Microsoft's Bill Gates had guns at Overture's head. Either one could say, "Take my offer, or I'll go to Google and your stock will tank. Then I'll buy you for pennies on the dollar." Wall Street understood this, and was trading Overture at a discount. What to do?

In early 2003, Overture made its move. In one week, Meisel and Gross bought the upstart AlltheWeb engine as well as the ailing AltaVista, gaining a broad portfolio of search patents (including Louis Monier's original work), as well as what might be considered a miniportal. AltaVista seemed perfect for Overture. The acquisition signaled that the company was willing to restore the AltaVista brand's original glory if the markets forced its hand. But as it stood, the site, with just 5 percent of the traffic brought in by Yahoo or Microsoft, was not a threat to Overture's partners. And by acquiring both the AlltheWeb and the AltaVista search technologies, Overture could claim to Wall Street that it had become a "full-service search solutions" company, able to compete with Google on both organic and paid listings.

But the AltaVista and AlltheWeb purchases were a hedge. At the end of the day, Overture had another route in mind: selling to one of its partners.

The Yahoo Deal

In another life, Ted Meisel must have been a poker player, but even the greatest players sometimes fold before the river card is turned.

Three days before his company signed a definitive agreement to be acquired by Yahoo, Meisel and I sat down for a chat in his Pasadena office. Overture was a fascinating story, but save for the occasional news item, it had been largely ignored by the mainstream business press. Despite its role as the largest pure play in the search field, on track to clear nearly $1 billion in revenues in 2003, it lacked Google's sex appeal and broad consumer brand.

After discussions with Overture's partners, advertisers, board members, and investment banking analysts, I had a few questions for Meisel. First among them: why is Overture an independent company? It was difficult to find anyone (besides Gross and Meisel) who thought Overture had a future as anything other than a division of either Microsoft or Yahoo. Its role as a behind-the-scenes paid listing provider meant it was dependent on Yahoo and MSN for nearly two-thirds of its revenues, and Wall Street had begun discounting its stock as a result. Industry wags had started to game its acquisition, and most had given the edge to Yahoo, which depended on Overture for 20 percent of its revenues and even more of its profits—clearly an untenable situation for Yahoo CEO Terry Semel.

Meisel says that his board had considered such a scenario, and decided "it doesn't fit within our view."

In the Internet media market, he continues, warming to the spin with the confident terminology of a practiced consultant, "you need a neutral party that executes well." He argues that vertical integration—

where a company owns everything from manufacturing to distribution—is not presumptive in any industry. In other mature industries, competitors have figured out the boundaries with their shared suppliers. Yahoo and Microsoft would do the same with Overture, which would remain neutral. "Automakers," he says by way of example, "don't compete with their suppliers."

Did this mean Overture was indeed destined to be Delco to Yahoo's GM and Microsoft's Daimler Chrysler? Meisel laughs, then changes the subject. Clearly there were other cards on the table that day.

The following Sunday, Overture showed its hand. Bill Gross called me late Sunday night and left a message on my voice mail: "Things are about to get very interesting." On Monday the news broke: Overture had agreed to a $1.63 billion acquisition by Yahoo, and the competitive landscape in the Internet media business clarified. Vertical integration may not be the Internet media industry's final structure, but it's certainly looking that way for now.

When the dust settled I called Gross back and asked him how he felt about selling his brainchild to Yahoo when, in effect, he could have competed with Yahoo. But Gross was far too smart to cry over spilled billions. "We did very well with the Overture sale," he reasons. "We had invested the first $200,000 to start the company, and we invested in later rounds as well." Gross pauses, then allows himself a shade of regret. "We didn't get all the value that we could have," he acknowledges, "and that is bittersweet. But it was definitely our most successful deal to date."

GoTo/Overture may be IdeaLab's greatest success to date, but any triumph Gross claims is overshadowed by what might have been. Gross saw the opportunity first and he built a world-class company to take advantage of it, but in the history of search, Overture will remain a footnote.

Perhaps that's why Bill Gross isn't finished dreaming the next great dream. His companies have sold for $1 million, then $10 million, then $100 million, and now more than a billion dollars, but he's still not satisfied. So what's he working on now? Well, first and foremost, there's a small start-up at IdeaLab that is doing desktop search—Magellan all over again. And there was a photo search company called Picasa, but he soon sold that to Google. Neither of those, however, was the really big idea. The really big thing is . . . well, I'll let Bill explain it.

"Basically I have the next paradigm in search," he tells me. "It's the next economic model and the next relevance model."

In the fall of 2004, Gross delivered his answer: SNAP, a new breed of search engine that ranks sites by factors such as how many times they have been clicked on by prior searchers, among many other things. And true to form, Gross is innovating in the business model: SNAP has developed a pay-for-performance scheme that goes pay-per-click one better: advertisers can sign up to pay only when a customer converts—in other words, when the customer actually buys a product or performs a specific action deemed valuable by the advertiser, like giving up an e-mail address or registering for more information.

What motivates Gross to start all over again? One word: Google. "The relevance is going down on Google—it's starting to falter, mainly because of the gaming." In other words, Google is getting spammed up, this time with sophisticated search engine marketing techniques and click fraud, just as AltaVista was destroyed by simplistic porn hacks back at the launch of GoTo. "I think I have a search engine spam solution. I think I got it," Gross tells me. "I think I can do it."

Chapter 6
Google 2000–2004
Zero to $3 Billion in Five Years

If you want the position of God, then accept the responsibility.
—Christopher Eccleston as the Son of God, via Orbital[1]

Near the end of 1999, Google Inc. had thirty-nine employees, most of whom were engineers of one stripe or another. Omid Kordestani, Google's newly hired sales chief, was still plowing the fields for enterprise deals, but they were few and far between. With more than $500,000 (and growing) going out the door each month and less than $20 million in the bank, you didn't need a Stanford PhD to do the math: the company needed a business model that worked.[2]

There was always the fallback of simply running banners on Google's prodigious traffic—one deal with DoubleClick, an ad network that specialized in serving graphical banners, would probably net the company millions of dollars. But that felt like a sellout—DoubleClick's ads were often gaudy and irrelevant. They represented everything Page and Brin felt was wrong with the Internet. "They didn't want to turn the Web site into the online version of Forty-second Street," recalls investor and director Michael Moritz.

Instead, the young executive team decided to try a more focused approach—it would sell text-only ads to sponsors targeting particular keywords. When you searched for "Ford cars," for example, an ad would appear at the top of the results for Ford Motor Company. These first advertisements were sold on a cost per thousand (CPM) model. In other words, the model was based on eyeballs—advertisers paid by the number of "impressions" Google delivered.

Despite the rise of Bill Gross's GoTo.com and its pay-per-click model, in early 2000 CPM was still the dominant business model for most types of advertising—including DoubleClick's. The distinctions Google's founders insisted on—that the ads be text only, and that they be targeted at a searcher's query—represented something of a last stand before Google fell back to the more familiar turf of Forty-second Street. "Our theory was, well, we'll try this for a little while," Brin says, recalling how he and Page made the decision to try targeted text ads. "But if we start to see that we're running out of money, well then we'll just turn on a deal with DoubleClick, and we'll be fine because we have a lot of traffic." Brin and Page were idealistic, to be sure, but not to the point of suicide.

Mixing CPM with keyword-based advertising results had proven somewhat successful at Kordestani's previous job at Netscape, but he was selling banner ads, not text links. No one had any idea if the text ads would work. At the end of 1999, Google began testing a hand-rolled version of its new system. In January 2000, Google's first paying customers appeared on the site.

Turns out the ads worked well enough, but they didn't scale. Revenue was limited by Kordestani's ability to sell, and despite his talents, it was difficult to book enough orders to create a healthy business. "It didn't generate much money," Brin recalls, referring to the program as a "hand-patched life preserver." DoubleClick, he adds, was the ocean liner Google would swim to should the life preserver fail.

As spring 2000 approached, it looked increasingly likely that Google would have to swim for it. But fate intervened: in March, the NASDAQ market crashed. Over the next few quarters, it continued what became a historic slide. Cash-rich technology companies began slashing their marketing spend, and their mainstream counterparts immediately followed suit. By the end of the year, advertising revenues across the media business had plummeted. In this environment, not only were customers for Google's new text-based advertising system few and far between; the notion that DoubleClick could somehow save the company was also called seriously into question. By the end of 2000, DoubleClick's stock had plummeted from a high of nearly $150 to a low of around $15.

"We always thought we could swim to the boat," Brin recalls with a laugh. "But there was no boat!"

Had the bubble not burst, Google might have adopted a more traditional approach to Internet advertising. But the crash of the banner advertising market and the meager revenues from Google's first attempt at text advertising led Brin and Page to turn their gaze toward GoTo.com. And as little as they might like to admit it, they saw salvation in Gross's approach.

Brin and Page "very adroitly and cleverly fastened on the proposition offered by GoTo," recalls Moritz, who as a board member of both Google and Yahoo got to see the dot-com wipeout from a particularly privileged vantage point. "Had Google not adopted some of the advertising techniques that were working for others, [it] would have ended up a small, but nice, high-end company."

Google essentially copied GoTo's approach, building an automated self-service model that allowed advertisers to buy text ads online with a credit card.[3] Unlike GoTo, Google already had plenty of traffic for its natural search results, and Brin and Page made a point of separating Google's advertising results from its natural search results, a key distinction from GoTo, which launched as a purely commercial engine (though it later adopted a similar church and state approach).

In October 2000, Google introduced its new service, which it called AdWords. An announcement on the main site promoted the new service: "Have a credit card and 5 minutes? Get your ad on Google today." Despite Google's fabled devotion to speed and economy (Brin and Page obsessively counted every word on the home page), the promotion stayed up in some form or another throughout most of the fall, demonstrating how critical this new revenue lifeline was to the young company.

Initial versions of AdWords maintained the CPM approach—advertisers still paid for impressions instead of clickthroughs. But despite that, the service was a hit—revenues began flowing in, and the mood improved significantly around the burgeoning Google campus.

Dealing with Growth

And burgeoning it was. Despite the revenue woes, Google as a consumer service was absolutely on fire. By August 1999, Google was serving 3 million search queries a day. In September, the company took the beta label off its service, introduced its now familiar logo and design, and launched GoogleScout, a feature that suggested related pages to visit based on the pages you found using Google.[4] The company announced it was serving 3.5 million searches a day—as many as 65 each second.

By mid-2000, searches per day had swelled to 18 million, and the Google index surpassed 1 billion documents—making it by far the largest search engine on the Web. (Google made plenty of public-relations hay out of the event, adding a McDonald's-like page count—"searching one billion pages"—on the home page). Much of Google's new traffic was due to a deal the company struck with Yahoo—the very deal that Moritz had foreshadowed when he made his initial investment. In June 2000, Google replaced Inktomi as Yahoo's core search service. Not only did the deal validate Google's technology and bring swarms of new users to Google's brand; it also brought a new investor: Yahoo purchased a $10 million equity stake in its new partner as part of the deal.

By the time AdWords made its debut at the end of 2000, Google was serving 60 million searches a day. Its business model may have been shaky, but Google was taking off as a brand, despite the fact that so far, the company hadn't spent a dime on marketing.

Not that Google hadn't thought about marketing. In May 1999, Brin persuaded Susan Wojcicki, his former landlord, to join Google as marketing manager. Brin and Page knew that Google needed a marketing strategy, but they weren't sure what it should be. Later that summer, the company signed Scott Epstein, a veteran Internet marketing executive, to a three-month contract as an interim vice president of marketing. The interim approach reflected Page and Brin's reservations about promoting Google; they were not convinced that traditional approaches to brand building were appropriate given the service's remarkable organic growth. But in the bubble mentality of 1999, everyone was spending money on branding. The Internet was viewed as virgin territory, and Get Big Fast competed with First to Market Wins for Internet slogan of the year.

Epstein and Wojcicki set about determining a strategy for the young company. They didn't get a lot of guidance from their bosses. "It wasn't clear what I was supposed to do," Wojcicki said. "Our competitors had huge marketing budgets—AltaVista was spending $120 million on marketing in 1999. I figured we needed a logo, so I started with that."

Wojcicki and Epstein considered hiring a major branding firm like TBWA\Chiat\Day to revamp the company image and advertising. "We talked to all the agencies and we spent a lot of time on it," Wojcicki said. "We were being rejected by a lot of ad firms at the time because nobody knew who we were. In fact, we would say, 'We're from Google,' and they'd look at our logo and say, 'Oh, is that a children's clothing company?' "

I asked Wojcicki what her goal was in hiring an ad agency in 1999. Was it to make Google a household term? "Yeah, it was," she admitted, then chuckled. Epstein had plenty of experience with large marketing budgets—he once served as director of marketing for Excite, a major portal that spent millions of dollars on marketing and branding.[5] Epstein brought in the Z Group, a marketing consultancy run by Sergio Zyman, the former head of marketing for Coca-Cola (infamous as the mastermind behind New Coke). Armed with consumer research from Z, Epstein presented a multimillion-dollar consumer marketing strategy to the founders and the board.

In the end, the founders' Burger King ethos prevailed. Epstein's contract was not renewed. Senior management—including Google's new board—nixed the initiative. "It was a hard decision to make," board member and early investor Ram Shriram recalls. "We were the only company not spending money on marketing. Were we the dumbest people in the business?"

"Marketing could have killed the company," Wojcicki reflects, "because we were going to spend like five or ten million dollars. We only had twenty million. Imagine, you cut us in half; suddenly we would have had to look for money or we would have had to do banner ads or something. We would not have had the luxury that we had later on."

By eschewing traditional approaches to marketing, Brin and Page were betting on a phenomenon that had proven reliable: that of public relations. Google was already a press favorite; glowing mentions of the company were coming in nearly every day. About the same time as Epstein was working up his marketing plan, Brin and Page hired Cindy McCaffrey, a veteran public relations executive, as director of corporate communications. She urged Brin and Page to adopt a "press first" approach to promotion. McCaffrey had helped guide Apple's press relations during the rise of the Macintosh in the late 1980s and she saw the same kind of buzz building around Google.

"Our approach became to invest in the product, and use PR as a tool for getting people to read and talk about Google," McCaffrey recalls. "Once they tried it, they'd like it. It became a turning point for Google."

A March 2000 article in *Time* magazine represents how McCaffrey's early strategy bore fruit. In the piece, headlined "Gaga Over Google," author Anita Hamilton gushed: "The great thing about Google is that it works. I had a feeling there was something different about Google when not one, not two, but three different friends recommended it to me."

With press like that, who needs a Super Bowl ad?

The Infrastructure Rules

As Arthur C. Clarke once observed, "Any sufficiently advanced technology is indistinguishable from magic." Google garnered impressive word of mouth among its users for one reason: it worked. Not only did its PageRank-based algorithms produce delightfully relevant results, but they did it with impressive speed, and the service never showed signs of buckling under the exponential growth it was experiencing.

Page and Brin had their Stanford-era frugality to thank for this robustness. Because the pair had to scrape for every machine they could find to support the early service, they were forced to optimize Google to run over off-the-shelf parts—cheap hard drives, cheap memory chips, and cheap CPUs. Instead of buying heavy mainframe artillery from the likes of IBM or Fujitsu, Brin and Page created a small army of foot soldiers—a massively parallel formation of cheap processing and storage. The beauty of the system was that it scaled—the more computers you threw at it, the more robust it became. And when a component broke down, no problem; you simply swapped it out. The system itself could never fail—there were simply too many individual parts, none of which depended entirely on the others.

This approach, known as distributed computing, would soon become all the rage in corporate environments. Even IBM realized its value, introducing a line of cheap servers it called blades in early 2002. But Google took it many steps further, developing its own operating system on top of its servers, and even customizing and patenting its approach to designing, cooling, and stacking its components. While nobody was paying much attention to Google's approach to computing back in 2000, this approach would become the company's core defensible asset by the time it was ready to go public in 2004. (Google's other major asset—the PageRank patent—is, in fact, owned by Stanford University, but licensed exclusively to Google until 2011).

Who Should Run Google?

As 2000 progressed, Google began to hire, slowly at first, but by the end of the year, the pace picked up considerably. Wojcicki was tasked with much of the hiring administration—the founders insisted on not using recruitment consultants, which were common in venture-funded start-ups.

"We learned early to do as much as we could in-house," said Wojcicki, who has gone on to become director of product management for the company. "That especially holds true for hiring."

The company went from a handful of employees to nearly forty in its first year; by the end of 2000, it had grown to nearly 150. It was during this early expansion that Google's unique approach to hiring became apparent. To say the founders obsessed about who might join the company was an understatement. Forged as they were in the start-up culture of Silicon Valley, and cognizant of the travails Page's brother and other friends were enduring with their own early-stage start-ups, Page and Brin were determined not to repeat their friends' mistakes. Prime among them was the hiring spiral.

In a hiring spiral scenario, the founders hire a person they might consider an A—perfect for the job, intelligent, productive, and a good cultural fit. They then let that person hire other people, and those new people hire more people, and so on. The problem is, A's often hire folks who don't threaten or challenge them—B's, to continue this rather Huxleian metaphor. Those B's repeat the pattern, hiring C's, and so on, until your company is quite literally consumed by C- and D-level people who are there for all the wrong reasons. The company loses its unique culture and falls victim to divisive internal politics and the malaise of hierarchically driven management games.

Page and Brin were not going to let that happen at Google, and to prevent it, they created hiring committees that reviewed every single open position. That way, it wasn't just one employee's opinion that gated a person joining the company; it was more of a pluralistic debate. In the early days, every employee interviewed each new potential hire, and the small staff argued for hours over who could or could not join the company. "I interviewed every single candidate for a job," recalls Shriram.

According to some early insiders, the hiring process felt like the rush process at an exclusive fraternity house. (This was not entirely accidental. Google executives still compare Google's internal culture to the collegial atmosphere of an elite graduate school.) As the new company continued to grow, the concept of hiring committees was expanded, with groups focused on various aspects of the business.

But the one hiring committee that mattered most—the board committee responsible for hiring the CEO to replace Larry Page—had yet to make any progress. And the venture investors were starting to get restless. The new AdWords program had bought the company some time, but it still wasn't making money, and the pressure was intensifying on the young founders to either make something happen or get out of the way.

"You have a balance between the natural impatience of an investor, and the nervousness of a founder about bringing in the CEO," recalls Moritz, choosing his words carefully. "You want to find some amenable middle ground. It is easy to make the wrong choice, and it's costly if you do."

Was there pressure from the investors on Page and Brin to find a replacement? "Yes," Moritz admits. Did it take longer than he would have liked? "It would be disingenuous if I didn't admit that," he replies. "This was a long and protracted process."

Over the course of eighteen months, from June 1999 through early 2001, Page and Brin reviewed more than seventy-five candidates for the CEO position. After several months it became clear that the founders were not impressed by any of the executives with marketing or sales backgrounds—they simply didn't speak the same language. Whoever might end up passing Brin and Page's test would clearly have to possess serious engineering chops, and would have to put up with their clearly demonstrated penchant for control. Of course, to pass muster with the investors, the ideal candidate would also need significant management and leadership skills. Finding someone with that combination of skills was proving extremely challenging.

Enter Eric Schmidt

In April 2004, Eric Schmidt returned to his alma mater, the engineering school at the University of California, Berkeley, to give a speech. Schmidt represented a major success for Berkeley, as the engineering school had long played second fiddle to Stanford in the hierarchy of prestige and funding. Sure, Berkeley was a good school (Sergey Brin had considered attending, but chose Stanford because it was, in his words, "cleaner"), but it seemed Stanford's graduates were the ones starting all the cool new companies, from Hewlett-Packard to Google. Stanford's perceived superiority in engineering was a small but significant aspect of a venerable and oft-contested rivalry between the two great universities—one public, the other private; one a bastion of messy liberalism, the other with a more buttoned-up and conservative bent.

So when Berkeley's School of Engineering welcomed Google CEO Eric Schmidt, PhD '82, back to campus for a guest lecture, Dean A. Richard Newton was in an ebullient mood. As he introduced Schmidt, who by then had been CEO of Google for nearly three years, he retold an old joke about Stanford engineers and their counterparts at Berkeley. "Many of you, like [me], probably were the butt of that joke we've heard for the past ten to fifteen years in Silicon Valley—'What do Berkeley engineers call Stanford engineers?' " The lunchtime crowd of alumni and faculty members laughed, then began to cheer, sensing what Newton was about to say next. "The answer in those days was 'boss,' " Newton continued. "I'm very pleased to say we've turned the tide on that, and Eric is the luminary that has set that standard." The crowd roared.

Were it truly that simple, Schmidt could have enjoyed that moment of homecoming appreciation at Berkeley, but one could detect a note of equivocation in his voice as he thanked Dean Newton and took the stage. Sure, Google was about go public in the largest IPO in Silicon Valley history, and sure, Schmidt was the CEO. But was he really the boss?

He certainly isn't the boss of Larry Page and Sergey Brin—the three share power in an unusual triumvirate structure that is based on consensus and partnership. Schimdt says he is comfortable with the agreement, but some close to him doubt that assertion. After all, they reason, it can't be easy to be CEO of the most successful public company in recent history, on the one hand, and yet be subject to the whims of two young founders who can outvote you two to one (and often have, according to various sources) on the other hand.

Critics of Google's structure, many of whom can be found, but few of whom will speak on the record, claim that Schmidt is simply a warm suit responsible for keeping Wall Street and the press happy, and that all major decisions are still made by Brin and Page. That the founders' fingerprints are all over the major decisions at Google is indisputable, but the role Schmidt plays in those decisions is more subtle than Google's critics might make it out to be.

Eric Schmidt comes across as a man who is comfortable in his own skin. He's been a CEO or top executive for more than two decades, having been CTO at Sun, where he made his first small fortune, then CEO of Novell, a major IT company, where he made his second. He knows when to smile, when to be gracious, when to keep quiet, and when to answer a difficult question with self-effacing acknowledgment. He brandishes subtle and humorous double entendres like a Japanese swordsman, a trait that almost offsets the superiority complex he shares with nearly every talented engineer in the Valley.

Despite these skills, one gets the impression that Eric Schimdt has yet to get entirely comfortable with his place at Google, his title as CEO notwithstanding. He's preternaturally calm, yet his demeanor feels slightly forced. To understand why, it's worth returning to 2001, when Schmidt was CEO of Novell and Google's CEO search was well into its second fruitless year.

"You will see nothing but 'wrong' when it comes to this story," Schmidt tells me, referring to his assumptions going into the process of becoming CEO at Google. "I had heard I was on the list," Schmidt says. "I thought that was pretty foolish. I thought search was not that interesting."

In early 2001, Schmidt fielded a call from Sergey Brin. Brin wasn't calling about the CEO position; instead he wanted to talk to Schmidt about Wayne Rosing, who was interviewing at Google for a senior engineering position. Rosing and Schmidt had worked closely together at Sun, and Brin was checking out Rosing's references. Schmidt figured the call wouldn't take very long, so it was scheduled for the end of the day, at 5 P.M. But the call took nearly an hour. "For a reference!" Schmidt recalls. "And from some kid? I thought that was odd. Just bizarre. I was trying to be helpful, but [Brin] was really, really going deep."

Toward the end of the call, Brin invited Schmidt to come to Google and meet Page and some others. Schmidt was noncommittal, sensing that Brin was feeling Schmidt out on the CEO position. But Schmidt's recruitment continued when his friend and Google board member John Doerr cornered Schmidt at a political fund-raiser a month or so later. Doerr, who has a record of getting what he wants when it comes to executive talent, asked Schmidt to accept Brin's offer of a visit.

"He said 'Why not just talk to them?' " Schmidt recalls. "I said,

'You have got to be kidding!' " But Doerr wore Schmidt down, and a few weeks later he found himself the CEO of a billion-dollar IT giant, sitting in an office with two twenty-seven-year-old kids whose business still lacked a proven revenue model. Brin and Page's approach to the interview only made the scene more surreal: on the wall of their shared office was a projection of Schmidt's biography, courtesy of Google's search service.

Google's chef—their *chef*?—brought in some food, and for the next hour and a half, the trio argued over just about everything. Page and Brin reserved their most withering attacks for Novell, the very business Schmidt was responsible for running.

"They criticized every single technical point I made, and everything I was doing in my business," Schmidt recalls with an odd kind of relish. "For example, [at Novell] we were building a series of caching proxies that would accelerate nodes within the fabric of the Internet. They argued that this was the stupidest thing they'd ever heard of—you wouldn't need it. I was just floored. It was just really arrogant."

Why Brin and Page chose Schmidt as their CEO after spending the better part of two hours denigrating his every move is an interesting question, but Schmidt hadn't been challenged like that for a very long time. He left Google impressed with the founders, and with the way they approached the interview process. "Of course, I thought I was right, and Larry and Sergey were wrong, but I made a note to myself, this is a pretty interesting company," he says. The founders were looking to test the new CEO against the same standard that Page and Brin used when they first met on the hilly streets of San Francisco—would Schmidt withstand the founders' intense and sometimes offensive style of intellectual fisticuffs?

Schmidt's answer to that question is interesting for its clarity: "Six months later I went back and checked [on the substance of the debates the three held that day in Page and Brin's office], and everything they said was right. That is kind of humbling—beat by two twenty-seven-year-olds."

Clearly the alpha dog hierarchy—typical of most engineering-driven companies—is alive and well at Google. But there were other reasons that Schmidt was interested in changing jobs. Running Novell was not exactly a picnic—Schmidt had been laboring for five years to turn the lumbering giant around. He had to restructure it to compete in the Internet age as well as with Microsoft, which had made a major push into the networking space. It was not a lot of fun, and the commute from his hometown in Silicon Valley to Salt Lake City, Utah (where Novell was based), was draining and demoralizing.

The prospect of a job next door in Mountain View—with a promising start-up backed by friendly VCs—had Schmidt intrigued. While AdWords had not yet taken root, it was looking healthier every week, and the company was offering him a sizable equity stake and the option to buy more should he care to. It all sounded pretty good, compared with being the public face of a public company that seemed on a slow and irreversible downward spiral.

Plus, Google was not competing with Microsoft, at least not yet. Schmidt had spent most of his career locked in a frustrating competition with Microsoft, first at Sun, which created an alternative platform to the Windows/Intel hegemony, and then at Novell, which owed much of its decline to Microsoft's entry into the networking marketplace. But search? Microsoft didn't have a dog in that fight.

Encouraged by Doerr and tired of running a large public company, Schmidt agreed to sign on at Google. "Big public company jobs are hard, and the satisfaction you get is in winning over a long period of time," Schimdt says. "I wanted to be closer to home, at someplace smaller and more manageable. And where the technology was more compelling."

But what of the lack of a proven business model? "I figured we'd sort it out," Schmidt says. "I told John [Doerr] that I'd give it a couple of years."

Schmidt eased into Google, announcing first that he would be leaving Novell as CEO in early March 2001. Schmidt was in the midst of completing a merger with Cambridge Technology Partners, an IT consultancy, and he had to stay on at Novell until the deal was closed. Schmidt joined Google in two steps, first succeeding Brin as chairman in March, and then taking Page's role as CEO three months later.

The industry response to Schmidt's new role fell out along the lines of either "It's about time Google got a grown-up onboard," or "What the hell is Schmidt thinking? Doesn't he know the Internet is over?!" After all, by the summer of 2001, the industry was in the throes of a devastating recession. But Schmidt's timing couldn't have been better—Google would claim its first quarter of net profits the very month he joined. And since then, the company has never had a down quarter. Either Schmidt was a genius, or he was very, very lucky.

Don't Be Evil

In July 2001, just a month after Schmidt joined the company, the triumvirate met to address what would become a fundamental challenge to the young company's future: how to manage growth. Google was already well past two hundred employees, and had moved from its University Avenue offices to new headquarters in a sterile but serviceable office park on Bayshore Parkway in Mountain View. But with all the changes, and all the new people (Google was hiring an average of five new employees a week), how might the company ensure that its original DNA—the founders' vision, values, and principles—remained intact?

The founders asked Stacy Sullivan, then head of Google's human resources, to round up a cross section of early employees with the mission of elucidating Google's core values—what was it about this place that made it special? How should Google employees treat each other? What are Google's core principles as a business and a place to work?

This particular brand of corporate soul-searching is typical for just about any young company experiencing hypergrowth, and it reflected

Page and Brin's very real concerns about avoiding the hiring spiral. Silicon Valley companies often become odd pastiches of the various cultures that preceded them—a clutch of ex-Netscape folks over here, a gaggle of former Apple folks over there. Instead of gelling into a new culture, growing companies can soon lose their identity as cliques develop that supersede the core values of the company itself. Identifying this problem and asking the head of HR to come up with a strategy to address it was nothing new. What was new, however, was what came out of that meeting.

On July 19, 2001, about a dozen early employees met to mull over the founders' directive. Joan Braddi, now vice president of search services, was there, as were David Krane, director of corporate communications; and Amit Patel, an engineer and employee number seven. Sullivan moderated the discussion, which began with the assembled group listing the core principles that they believed represented what Google was all about. The meeting soon became cluttered with the kind of easy and safe corporate clichés that everyone can support, but that carry little impact: Treat Everyone with Respect, for example, or Be on Time for Meetings.

The engineers in the room were rolling their eyes. Patel recalls: "Some of us were very anticorporate, and we didn't like the idea of all these specific rules. And engineers in general like efficiency—there had to be a way to say all these things in one statement, as opposed to being so specific."

That's when Paul Buchheit, another engineer in the group, blurted out what would become the most important three words in Google's corporate history. "Paul said, 'All of these things can be covered by just saying, Don't Be Evil,'" Patel recalls. "And it just kind of stuck."

It more than stuck; it became a cultural rallying call at Google, initially for how Googlers should treat each other, but quickly for how Google should behave in the world as well. It helped that in the months after the meeting, Patel scribbled "Don't Be Evil" in the corner of nearly every whiteboard in the company. For an organization consisting mainly of engineers, whiteboards served as the corporate equivalent of the water cooler. The message spread, and it was embraced, especially by Page and Brin. "The phrase captured what we all inherently felt was already true about the company," Krane recalls. "It was the lyrics written over a melody that already existed."

"I think it's much better than Be Good or something," Page jokes. "When you are making decisions, it causes you to think. I think that's good."

But what happens when those decisions have to do with whether or not to do business by the rules of the Chinese government, or whether to allow the U.S. government to track the search histories of thousands of Americans?

Defining evil seems pretty simple when you're sitting in a conference room of a small but growing Internet company in 2001. But had that small group of early employees understood the standard it was creating for Google through the adoption of that motto, it might have reconsidered its support for the phrase. Don't Be Evil is a wonderful sentiment for describing the ethical boundaries of internal company dealings, but when your business is understood to be a global arbiter of human knowledge and commerce, sticking to such a principled stand can become extremely . . . *tricky*.

Not to mention that it smacked of arrogance—who were these Googlers anyway, and what right did they have to determine what was evil and what was good?

I asked Amazon CEO (and Google investor) Jeff Bezos if Google's motto rang true with him. His reply aptly sums up the reaction of many observers: "Well, of course, you shouldn't be evil," he tells me. "But then again, you shouldn't have to brag about it either."

Google Gets Big

The year 2001 stands as a pivotal point in the history of the Internet: the year the bottom fell out, on the one hand, and the year the medium found its footing and began to grow in a truly profitable manner, on the other. And in Google's brief history, 2001 stands as the year Google got big, in nearly every sense of the word.

By the time Schmidt joined, Google was handling more than 100 million searches a day. Early in the year, the company began a raft of significant improvements to its search service, starting with the purchase of DejaNews, a failed attempt at making money from Usenet, a public messaging system composed of more than 500 million discrete postings on nearly every subject imaginable. While the acquisition of such a data-rich asset went largely unnoticed, the move marked a significant departure for the company. By acquiring Usenet and adding it to the index, Google was actively seeking out new information, as opposed to passively spidering the Web. The move was consistent with what would become the company's new mission statement: "To organize the world's information and make it universally accessible and useful."

Google would continue this trend through 2003 and 2004 with the acquisition of Blogger, Picasa (a photo-sharing service), and Keyhole (a massive satellite imaging company), and the launch of Google Print. But it was during 2001 that Google's appetite for data began in earnest. The service added public phone-book information to its index as well as a new image search tool, complete with 250 million images. By the end of the year, Google's burgeoning index comprised more than 3 billion documents. At the same time, the company aggressively expanded internationally—by early 2002, it was serving search queries in more than forty languages. And 2001 saw Google's aggressive entry into the mobile market through partnerships with major players like Cingular, AT&T, and Handspring.

Clearly, Google was metastatizing—everywhere there was opportunity, it seemed the company was expanding. Google soon had more than one hundred engineers in the company, but no focused approach to managing how their time was spent. Unsure of the best way to handle such growth, the triumvirate set up a traditional management structure based on hierarchy—teams of engineers reporting to more than a dozen engineering managers, who in turn reported to Brin and Page. But the approach began to feel top-heavy and bureaucratic—it was slowing down innovation. In September 2001, Brin and Page gathered all the engineering managers together at a companywide meeting—then informed them they were out of a job. Most got jobs in other places in the company, but the founders had made a declaration—not only were they in charge, but things would be done differently at Google.

Instead of unwieldy, top-down projects that harnessed dozens of engineering resources, Brin and Page created a more dynamic structure in which small teams of engineers tackled hundreds of projects, all at once. Brin, Page, and other senior managers reviewed each project on a regular basis, and the best projects received further funding and human resources. A Top 100 list was soon developed, and engineers competed to make it up to a top ranking—not unlike Google's search results. The company launched Google Labs, where interesting new projects—the best of the Top 100—could have an early public preview.

This let-a-thousand-flowers-bloom approach to management was generally well liked inside the company, but it also rankled quite a few employees. "It became a very political place," says one senior engineer who is no longer with the company. Like nearly everyone who spoke with me frankly about Google, he requested anonymity. "Nobody had the authority to do anything without Larry and Sergey's approval."

The idea of company founders being unwilling—or unable—to give up power is not new. In fact, it's so common in Silicon Valley that it's got a name: entrepreneur's syndrome. But while Page and Brin's unique approach to management angered some, others blossomed under it, and the company certainly continued to innovate.[6]

It would have to—the competition was growing fierce. With the growth of AdWords, Google's 2001 revenues were on pace to hit nearly $85 million. But Overture was growing faster—its 2001 revenues stood at a whopping $288 million. Overture was making a habit of exceeding Wall Street's expectations, and when it turned profitable, it did so with a bang. Net income for the fourth quarter of 2001 alone was $20 million—nearly one-fourth of Google's entire revenue base.

Google executives certainly took note of Overture's success, and it was not hard to figure out why the business worked: its auction-based pay-per-click advertising network had tens of thousands of clients. By comparison, Google's AdWords product was far less robust—throughout 2001, it still depended on a CPM model. The lack of both an auction and a pay-per-click component seemed to be limiting the network's growth.

It wouldn't be long before Google fixed those shortcomings, adding an important twist in the process. In February 2002, the company launched a new version of AdWords that included auction and pay-per-click features, but with this service—unlike Overture's—advertisers couldn't just buy their way to the top listing. Instead, Google incorporated the notion of an ad's popularity—its clickthrough rate—into its overall ranking.

This shift was simple, brilliant, and extremely effective. Imagine that three accounting firms are competing for the right to target their ads to the keyword "accounting services." And assume further that Accountant One is willing to pay $1.00 per click, Accountant Two $1.25, and Accountant Three $1.50. On Overture's service, Accountant Three would be listed first, followed by Accountant Two, and so on. The same would be true on Google's service, but only until the service has enough time to monitor clickthrough rates for all three ads. If Accountant One, who paid $1.00 per click, was drawing more clickthroughs than Accountant Three, then Accountant One would graduate to the top spot, despite his lower bid. Industry observers quickly dubbed the new approach AdRank, after Google's famous PageRank algorithm.

Google's decision to factor clickthrough into an advertiser's ranking forced an economy of relevance and profit into the pay-per-click model—after all, if the $1.00 merchant is generating five times the clickthrough of the $1.50 merchant, it only makes economic sense to give the $1.00 merchant the top spot—he's making Google, which gets a percentage of every click, more money. But the press and industry didn't see it that way—instead, Google was credited with being "less evil" than Overture, because it was not allowing advertisers to simply buy their way to the top of the advertising heap. It was yet another example of the Google PR halo at work—Google was the little company that had only the best interests of the users at heart, and by not being evil, it was rewarded with glowing press mentions and increased business from advertisers.

Google News

On September 11, 2001, just about everyone on the planet realized that the world had changed. Of course, everyone with a television had it tuned to a news channel, but that wasn't enough. Hungry to comprehend those cataclysmic events, much of the wired Western world turned to the Internet, overwhelming the servers at cnn.com, abcnews.com, and ap.com. Starved of consequential information, millions of Internet users took matters into their own hands. Unable to access traditional news sites, they turned to Google, flooding the servers with queries like "Osama bin Laden," "Nostradamus," and "World Trade Center." The world had just changed, and Google's users expected the service to help them understand how.

News-related searches on Google increased by a factor of sixty the week following the attacks, according to an academic paper on how Google responded to the events.[7] But the amount of traffic that hit Google in the weeks after 9/11 was about the same as before—by the end of 2001, Google was already serving nearly 125 million queries a day. Google's searchers simply changed what they were seeking from "Hank the Angry Dwarf" and "Britney Spears" to "World Trade Center" and "Afghanistan."

Google responded to the shift in interest with its first major editorial product—a news service that allowed its users to find and read copies of stories that were otherwise unobtainable owing to traffic loads on other sites. Because of its prodigious and scalable network architecture, for the first few days after 9/11, Google became the world's news service. Students of where Google might go in the future would be wise to recall this fact. The events of 9/11 taught Google and the world that Google had more than a search service at its disposal; it had an extraordinary asset—the ability to cache any information, at any time, and show it to anyone on demand.

Did Schmidt understand this when he joined? Of course he did. "Google has one of the largest data centers in the world, and one of the largest collections of bandwidth in the world," he tells me when I ask him to describe what he considers Google's core asset. "I get to ask, 'What would you like to do with it? What are the technological possibilities of that platform?' "

September 11 pointed the way to one new service that leveraged Google's core assets: Google News. Initially launched as a 9/11-related link at the bottom of the home page, by mid-2002 Google News had blossomed into a major hit. With search for images, a directory based on the popular Open Directory Project, and now news, it was clear that Google needed a new approach to communicating its burgeoning options. Concurrent with the News launch, Google redesigned its home page, adopting the familiar tabbed design now common to nearly every popular search engine. By 2002, Google as we know it had taken shape.

A Lava Lamp in Every Alcove

In May 2002, just months after it unveiled its new and improved AdWords product, Google announced its landmark deal with AOL. Not only would AOL begin employing Google's search technology; it would also be using Google's paid listings. In essence, Google was entering a new line of business: AdWords syndication.

This was the very line of business (and deal) that fueled Overture's initial growth. Lines were clearly being drawn: after losing

AOL, Overture managed Yahoo's and Microsoft's paid listings, while Google now had its own site as well as AOL's. Both companies had scores of lesser deals as well—Google powered Ask Jeeves and Earthlink, for example.

But the AOL deal was a major risk for Google. While its search technology was robust and capable of handling tens of millions of additional queries, the same could not be said of AdWords. Plus, to win the deal, Google had to guarantee AOL tens of millions of dollars in revenues (as well as a minor but valuable equity stake). What if the AdWords system had the equivalent of a recession, and keyword prices plummeted? If that were to happen, Google's debt to AOL could have forced the young company out of business.

"The AOL deal was a really big bet for our company," Brin tells me. "We thought it might bankrupt us. We had very little experience; it required a degree of growth. . . . I don't know what would have happened if we hadn't won that AOL deal."

But win it they did, and despite early concerns, the deal proved lucrative for both parties. The alliance with AOL shot Google into the A-list of major Internet players, alongside Yahoo, eBay, and Amazon, the survivors—and thrivers—in a new Internet age. A fresh round of press interest swamped the company, as did incoming queries from nearly every conceivable business partner, advertising client, and potential recruit.

By mid-2002, Google was on a tear. "No one can write a story about the Internet without 'Google' in the title," complained Steve Berkowitz, CEO of Google rival (and partner) Ask Jeeves, echoing comments I heard from nearly every other major competitor in search. It seemed the company could do no wrong—the press was in love with it,[8] and its users were rabidly loyal. Google had the highest loyalty of any online brand, according to a study done at the time by brand consultancy InterBrand.

To be sure, Google had the story everyone wanted to read. The company maintained the wacky nonconformity of the late 1990s, but coupled it with a do-no-evil, make-money-honestly philosophy consistent with a post–9/11, post-Enron business world. It was a perfect feel-good story.

And Google employees certainly felt good. They were quite proud of their benefits, the very same perks that had become symbols of Internet-era excess after the crash. They were explained as not simple excess, but, in fact, subtle and important recruitment tools. Geeks tend to be antisocial, the line went, and they need help with socializing. Hence, Google had a lot of parties and encouraged its employees to play while at work—that's why Google had volleyball courts, free scooters to zip around campus, and foosball and ping-pong tables in every building.[9]

Google's employees played up this advantage wherever they went. Google engineer Amit Singhal went so far as to include pictures of these perks in a presentation to a group of engineers at IBM. Not only did they get to see an overview of how Google works; they also got to see a photograph of Google's chef and game rooms.

While such a display might have motivated some IBM engineers to apply for jobs at Google, chances are it alienated a few as well. After all, Google didn't invent the freewheeling geek culture it espoused—it was simply the only company in late 2002 that was capable of affording it. To some, the presentation smacked of triumphalism.

Just Who Did These Kids Think They Were?

There are serious drawbacks to being the hottest company on the planet. As more and more people tell you that you can do no wrong, and as more and more profits, kudos, and recognition come your way, a company can begin to develop a culture of insular arrogance. By late 2002 and into 2003, it was clear that Google was developing a serious problem along these very lines. A backlash began to grow among the Silicon Valley elite, built on envy and jealousy, to be sure, but also on countless interactions with the company that left non-

Googlers with the feeling that Google was unresponsive, self-centered, and dangerously cocky.

"Google is going to have a major fall in the next couple of years," a well-known venture capitalist—one who did not get a piece of Google's deal—told me in early 2003. Echoing scores of private conversations, he added: "They've pissed off too many people."

"Some of their hubris is warranted," a major Wall Street analyst countered, before continuing: "But this cult of genius is going to be difficult to take out of the company."

It's worth picking apart that "cult of genius" sentiment, as it reflects a deeper set of circumstances that held true at the time. By mid-2002, the Valley was in its second full year of recession. Tens of thousands of young technology workers were unemployed, and no one was hiring. No one, that is, save Google. While the rest of the Valley languished, Google prospered.

A job interview at Google was widely viewed as the equivalent of a golden ticket to Willy Wonka's chocolate factory—the one fabled place in the Valley where time had stopped; where the lava lamps still glowed with the promise and optimism of the dot-com boom; where lunch was free, employee perks were legendary, and everyone was happy, healthy, and, should the company go public, quite rich.

Thousands of résumés streamed into Google each week, swamping the company's hiring process. Legions of talented geeks never got so much as an acknowledgment of their desire to work at Google. Hundreds of others got interviews but were never hired, and many of those felt spurned by a fickle and mysterious process that no one seemed capable of explaining. When hundreds of smart people feel poorly treated, the negative buzz starts to build. "A lot of [those we passed over] were certainly good enough, and it's something I regret," Brin acknowledges when I ask him about Google's hiring practices in late 2004. "It's something we have to fix."

It didn't help that, like many fast-growing tech companies before it, Google hired legions of full-time contractors, folks who worked just as hard as employees but did not get to go to company-wide meetings or Google's lavish holiday parties.

But it was not just hundreds of spurned geeks who began to bad-mouth Google; it was thousands of advertisers as well. By 2003, Google had amassed more than 100,000 advertisers using AdWords, yet its investment in customer service was minimal—it preferred to automate customer interactions. "We feel if it can be automated, it will be automated," Omid Kordestani tells me. This left many advertisers cold, and fostered even more ill will. Message boards populated by advertisers began to regularly bash Google for its seeming indifference to their issues and its apparently unslakable thirst for more and more control of the search market, and by extension, the entire e-commerce world.

Observers of Silicon Valley culture took note, and by the end of 2002, they began to view Google not just as a search engine with a neat culture and an impressive business model, but as quite possibly this generation's next great monopolist—first IBM, then Microsoft, and now Google.

The Valley wanted to connect to the burgeoning company, both to bask in its good karma and to reap the rewards of potentially lucrative partnerships. Folks called, e-mailed, and stopped by Google, but the overwhelming sense was that, from 2002 to early 2004, Google was simply not in the mood to listen, or to take advice, past a polite "huh" and an occasional "We'll look into that."

Why? Two factors come to mind. One, the company was terrified of messing up a good thing, and nearly paralyzed by its own success. Nearly all those I met with during that period acted as if their hair were on fire—too much to do, and far too little time to do it. Marissa Mayer, an original product manager at Google and a crucial cultural force in the company, is a good example of this. Mayer, a hummingbird of a woman who speaks faster than most humans can hear, will fly only on red-eyes—planes that travel at night. I asked her why. Her answer: she doesn't want to miss a single workday.[10]

The second factor comes down to the founders' characters. The company's founders are, upon first impression, strikingly similar to the persona that Google projected during those two years—aloof, supersmart, dismissive of unsolicited advice. They are, after all, first and foremost engineers. And engineers are not the best communicators, nor do they make the best diplomats or business development executives. They tend to trust technology before human beings, leading to a culture of limited information sharing. Many of the senior execs at Google operate with "an alienating and unnecessary secrecy and isolation," says Doug Cutting, a veteran Valley engineer who founded the open-source Nutch search engine.

True enough, but certainly nothing new. The same could be said of nearly every entrepreneur who tried something new and was rewarded with unimaginable fame and fortune.

In July 2002, Paul Ford, a well-respected observer of Internet culture, published a work of fiction on his weblog. Titled "August 2009: How Google Beat Amazon and eBay to the Semantic Web," the article laid out a compelling scenario for how Google could grow to control pretty much the entire online world. Ford illustrated the article with a crude doodle that showed the Googlebot—Google's indexing program—as a monstrous robot standing atop the world. "I am Googlebot," the cartoon declares. "I control Earth."

While the piece was a rather insightful and detailed explanation of how the semantic Web might work (for more on that, see Chapter 11), wags in the Valley latched onto Ford's characterization of Google as a technological juggernaut. Ford had hit a nerve, and not just with people outside the company. Oddly enough, many Googlers saw Ford's not-so-subtle put-down as validation of their superior position in the world. "Google marketing called and asked if they could put my rendition of the Googlebot on T-shirts for some sort of developers' summit," Ford tells me. "I figured, what the hell, and I sent them the image. But I made sure that Google folks would understand that the rights to the image belonged to me, and I asked for payment in shirts."

"That seemed to dampen their excitement," Ford continues. "I got the sense that Google doesn't like to be told what they can or can't do. That I might want something in return for my work made no sense to them."

Since that exchange, Ford learned from friends that his illustration was "liberally reproduced within Google . . . never with permission, or payment, of course. I guess Don't Be Evil doesn't apply to respecting copyright law."

The T-shirts, however, never did get made.

Google's venture backers noticed the cracks in their prized investment's facade, and concluded that the trio of Schmidt, Page, and Brin needed shoring up. The combination of the founders' strong wills and Schmidt's deference to the original culture meant that key management decisions were not being made, or if they were, they were not being made properly. Board member John Doerr asked that the trio let Bill Campbell, founder of Intuit and revered Valley veteran, come in for some informal "coaching" of the team. To its credit, the triumvirate agreed. Campbell began spending a few hours a week at Google. "God bless that man," Doerr told John Heilemann in an article penned for *GQ* magazine in early 2005. "I don't know where the company would be without him."

Google Marches On

Regardless of the growing backlash, Google was simply too big, and too good, for the naysayers to overcome. By the end of 2002, Google stopped publicly discussing its key internal metrics, claiming that it had "more than 1,000" employees and "more than 10,000" computers in its vaunted infrastructure. The company did still boast about the size of its Web index, which passed 4 billion documents in December 2002. But it guarded its revenue numbers jealously—perhaps because they were so good: in 2002, the company made nearly $100 million on gross revenues of about $440 million. That's some serious cash, and the longer people like Bill Gates stayed in the dark about it, the longer Google could remain free from additional competition.

As compared with Google the service, it has always been difficult to extract information from Google the company—clearly this trait was inherited from its founders, Page in particular. But in late 2002 and early 2003, it seemed the company was circling its wagons even more, perhaps for competitive advantage, but perhaps also in preparation for a possible IPO.

In December 2002, the company launched Froogle, an e-commerce search engine. To most, it was increasingly clear that Google planned to play, and big, in the world of e-commerce. Through the next year, the company continued its aggressive expansion and its rather disingenuous practice of avoiding hard numbers. In mid-2003, the company announced it served "more than 250 million queries a day," and as of early 2005, it has not updated the figure. In early 2003, Google acquired Blogger, the wildly popular weblog hosting company, prompting many to speculate that Google was becoming a portal along the lines of Yahoo or AOL. But Google for the most part left Blogger alone.

Why? The answer most likely lies in the company's next major innovation, a new advertising program called AdSense. Launched in

March 2003 and rolled out to the world that June, Google's AdSense program marked a departure in the company's business model—this was not a pure search business; it was something else. AdSense allowed third-party publishers large and small to access Google's massive network of advertisers on a self-serve basis—in minutes, publishers could sign up for AdSense, and AdSense would then scan the publishers' Web sites and place contextually relevant ads next to the content, much as AdWords did for Google's own site.

But there was a significant difference to AdSense—it was driven not by the intent-based queries of consumers, as search is, but rather by the content of a site. The presumption was that if a reader was visiting a site written about, say, flowers, advertisements about flowers from Google's networks would be a good fit.

By nearly any measure AdSense was a hit—thousands of publishers signed up for the service, most of them tiny sites that previously had no way to monetize the small amount of traffic they had garnered. This was particularly true for blogs—the connection to Blogger now became obvious. For many, AdSense was the equivalent of magic—they added a few lines of code to their sites, and in a month or so checks from Google started showing up in the mail.

But while AdSense as a revenue stream has grown steadily—by early 2005 it accounted for an estimated 15 percent of Google's overall revenues—many advertisers complained that AdSense didn't work nearly as well as AdWords. Potential customers are in a very different frame of mind when they are *reading about* flowers from when they are typing "flowers" into a search engine. Google acquiesced to advertiser feedback and in 2004, allowed them to opt out of the AdSense network. Regardless, AdSense was a major new distribution network for what can be considered Google's second most impressive asset, after its core infrastructure: its network of advertisers.

Chapter 7
The Search Economy

The last bastion of unaccountable spending in corporate America.

—Google CEO Eric Schmidt
on corporate marketing budgets

Neil Moncrief couldn't afford to have a bad quarter. In fact, even a bad month made things a bit tense at home—running your own business is like that. When things go south at the office, you take it home with you. "It's not like working for the man, where you leave it at the office at five o'clock," he tells me in a soft southern drawl.

As a small businessman, Moncrief lives on the edge of profit and loss—a bad month means avoiding his local banker, putting off home and car payments, and having less meat on his family's table. But Moncrief is proud of what he has achieved. He built a small e-commerce company, survived the nuclear winter of 2001–2002, and emerged with enough cash flow to take care of his family.

Moncrief has search engines to thank for that cash flow, Google in particular. Thanks to the traffic that Google drove to Moncrief's online storefront, Moncrief no longer worked for the man. But as the holidays approached in the year 2003, Moncrief got a new boss.

His name was Google, and he made old Ebenezer Scrooge look like a saint.

In mid-November, Google started messing with Neil Moncreif's business. Traffic to his site shriveled, cash flow plummeted, and Moncrief fell late on his loan payments. He began avoiding his UPS man, because he couldn't pay the bill. His family life deteriorated. And as far as Moncrief could tell, it was all Google's fault.

Moncrief is one of the tens of thousands of merchants who have taken to the Web since the Internet became a global phenomenon. For every household brand built during the bubble's infamous glory—eBay, Amazon, Expedia—thousands of Neil Moncriefs toiled in relative obscurity, building the Web's bike shops and insurance agencies, its button merchants and stroller stores. These digital cousins of strip mall America are the very beating heart of the U.S. economy—small business, writ large across cyberspace. You think Amazon's got scale? You think eBay is huge? Mere drops in the bucket. Amazon's 2000 revenues were around $2.76 billion. But the Neil Moncreifs of the world, taken together, drove more than $25 billion across the Net that same year, according to U.S. government figures. That's the power of the Internet: it's a beast with a very, very long tail. The head—eBay, Amazon, Yahoo—may get all the attention, but the real story is in the tail. That's where Moncrief lives.[1]

Moncrief's little piece of that tail is in shoes, in particular, big shoes. His company starts at size thirteen and goes up from there. Moncrief's a size fourteen, and as all in the fraternity of the large footed know, it's a pain in the ass to find shoes that fit properly. So Moncrief hooked up with a technically inclined friend "who handles anything with wires coming out of it," and the two launched 2bigfeet.com in 1999.

Moncrief's idea to set up shop online was pretty simple, and at the time, not particularly new. In fact, from 1995 to 2000, tens of thousands of business owners took out small-business loans from their local banks or the government in order to open storefronts on the Web.

Like all sensible and enterprising pioneers, Moncrief saw a new frontier, and he decided set up shop on it. The logic of selling big shoes over the Internet is quite compelling. Only a small percent of folks are big-footed, and they don't tend to hang out in geographically concentrated areas. Launching a chain of retail storefronts for such a thinly spread population would be a pretty huge waste of money. Moncrief does have one storefront in Georgia, but it's mainly a stockroom for the four thousand or so pairs of shoes he ships around the world every month.

It's fair to say that 2bigfeet.com is a business that owes its existence entirely to the geography-busting elements of the World Wide Web. On the Web, no one cares if you're based in Albany, Georgia. Folks in search of a decent-looking pair of shoes for their oversize feet are a pretty motivated set of customers. These are customers that, given the right tools, will search for your business, as opposed to making you search for them.

But while the Web may offer access to hundreds of millions of customers, you still have to let them know you exist. Back in 1999, there weren't a lot of options available to a small partnership with a few $10,000 small-business loans and a Web site. Moncrief couldn't afford to cut a big deal for real estate on AOL or Yahoo; he couldn't even afford mere banner ads on those sites. (Moncrief was suspicious of them in any case; he didn't believe folks paid them much attention.)

Given that he had no choice, Moncrief counted on the one thing he thought was a hard and fast rule in the Internet. When folks went looking for something, they usually started at a search engine. And through some combination of luck, good karma, and what seemed like fair play, when folks punched "big feet" or similar keywords into Google, Neil's site came up first.

Google Giveth, Google Taketh Away

Thanks to Google, the orders flowed in. Life was good. Sales took off, and soon Moncrief had a bustling business on his hands. He had done just about everything right—he found a need and he filled it. By the middle of 2003, Moncrief was moving more than $40,000 worth of big shoes a month, with 95 percent of it coming in through search engine referrals—the majority of those from Google. The best part: Moncrief had never purchased an advertisement—all those search engine referrals were "organic." People found Moncrief through Google because, well, Google worked as it was supposed to work. "I figured folks who had to buy an ad, well, there must be a reason they needed to," Moncrief told me. "We were the right answer for the search, so why buy an ad?"

Then, right before the critical holiday shopping season, a hurricane hit 2bigfeet.com.

In the third week of November—November 14, 2003, to be precise—the phone stopped ringing and the orders stopped coming in. For two weeks, Neil Moncrief didn't know what had hit him. But then he began to wonder—maybe Google was broken?

The very thought seemed ludicrous—*Google,* broken? But a quick search on Google confirmed his suspicions—2bigfeet.com was no longer the first result for "big feet" on Google. In fact, it wasn't even in the first hundred results. As Moncrief put it, it was as if the Georgia Department of Transportation had taken all the road signs away in the dead of night, and his customers could no longer figure out how to drive to his store. What the hell had happened?

In short, Google had tweaked its search result algorithms, something the company does quite frequently. But this time Google's modifications, which were intended to foil search engine spammers, had somehow sideswiped Moncrief's site as well. What Google giveth, Moncrief learned the hard way, Google can also take away.

Thanksgiving was looming, and Moncrief was facing the loss of his entire Christmas season. What to do? He quickly went to the Google Web site and attempted to find a number to call or an e-mail contact where he could petition for redress. After all, everything was working before, so why change it now? Why would Google, a billion-dollar Silicon Valley giant, take the time to single out a father of two who runs a tiny shoe business in Georgia? Doesn't Google realize, Neil wondered, that it's wiped out my business, my livelihood?

Well, in fact, no. Moncrief called Google's headquarters in Mountain View, California, but never got more than voice mail and nary a single call back. He e-mailed help@google.com and searchquality@google.com, but never got a response. It was as if the geeks out in California simply didn't care: they were leaving Moncrief twisting in the wind.

It was then Moncrief realized that while he may have stopped working for the man, he was now working for a far more capricious overlord, one who had no idea he even existed.

Moncrief is a cautious man, conservative, a Republican. He's not the type to ask for government intervention. But when I first spoke to him back in 2003, he was ready to string up the bastards at Google. They were messing with his family, he wasn't sleeping well at night, and *they didn't even return a simple phone call.* Neil had four or five bank loan books sitting on his desk, mocking him, and no money to pay them. And there were four thousand pairs of oversized shoes on his shelves, going nowhere but out of style.

How had it come to this?

The Google Dance

Neil wasn't the only one wondering. In fact, beginning on November 14, an entire industry of search fanatics went on full-blown alert, tearing up Internet message boards with speculation about the latest Google Dance—the moniker given to Google's periodic updates of its algorithms. These updates had grown increasingly dramatic, and this latest one, coming on the heels of a slew of hurricanes that had hammered the Sunshine State, was dubbed Florida by the search industry. It was Google's most dramatic yet.

On WebmasterWorld, the king of all forums for practitioners of search-engine marketing and optimization (SEM/SEO), the reports came in from around the world: Google was updating again, and this time, the SEO industry appeared to be the target. Google was directly filtering out some of the very optimization practices that had made the industry possible.

Because Google had become the source of so much traffic for so many, any burp or shudder in the company's indexes had exponential implications throughout the young world of search-dependant online businesses.

"Well, this is just terrific. *£%$"*@*¬!" posted a typically exasperated search-engine consultant. "I'm going have a simply fantastic day come Monday, explaining to clients why almost all their sites appear to have been removed from Google. GRRR! Why don't I just pack up shop now?! I'm going to get crucified. I cannot believe Google has done this again. Geez this makes me angry! (and severely worried about the future of SEO as a viable business)."

In short, Google had updated its indexes to penalize what the company viewed as spam—people gaming their sites so they ranked higher. And a lot of folks, including Neil, were caught in the cross fire. Neil was an unfortunate casualty of a much larger war, an arms race of sorts fought over relevance, power, and money.

GoogleGuy, an anonymous forum participant who works at Google and has the thankless task of damage control during updates, responded on the same day:

"Hey everyone, we're always looking for ways to improve the quality of our rankings and algorithms. I'll post more over the next few days—just wanted to let people know that I'll be around."

But while GoogleGuy did keep reading the forum, and even posted carefully worded exhortations that everyone should be patient, the net result of Florida was clear: Google had taken a major stand against what it determined were search-engine spammers, and those who felt their legitimate businesses got hurt were told, in essence, to pound sand.

As an agonized poster to the WebmasterWorld forum wrote, summarizing the complaints of thousands:

GoogleGuy please listen to what people are saying. A lot of us are hurting after this update. It couldn't have come at a worse time, just as the Christmas business was starting. Fifty percent of my business is gone overnight and I may need to lay off warehouse staff or have them standing about with nothing to do. We have always done everything by the book, I can only conclude that our large affiliate network has been penalized by the new algo. Overnight the bulk of our best affiliates have just disappeared out of the index together with two of our best performing sites.

Three years' hard work wiped out in 24 hours.

That may as well have been Moncrief posting, although Moncrief had never heard of WebmasterWorld, search engine optimization, or affiliates. When I asked him if he had engaged in any spamlike optimization practices, he threw up his hands. "I just have a site that sells shoes," he told me. "I'm not optimizing anything."

The SEO World

At this point it makes sense to step back and explain a bit about the SEO industry, and the affiliate spammers in particular. SEO grew out of the simple observation that being listed in the top few results on Google translated directly into cash. Look what it did for Moncrief: he built a significant business in oversize shoes with little or no marketing.

Still in its early days, the SEO business has a whiff of the Wild West about it. While most SEOs are legitimate businesses, many sites promoting optimization—the practice of tuning a Web site to rank better in organic search results—sport loud come-ons reminiscent of late-night television, replete with garish promises, many written in poor English (for some reason, SEO seems to flourish in Eastern Europe). It seems that many SEO practitioners share the same genes as hawkers of Ginzu knives, miracle vegetable juicers, and Ponzi schemes.

Wherever there is easy money to be made or an opportunity to game a system for profit, you'll find a fair measure of hucksters, cheats, and opportunists. Of course, you'll also find honest businesspeople making real livings. But back in 1999–2001, as Overture and Google began to provide a new business model for marketers and a fresh stream of what seemed like limitless cash for well-positioned Web sites, entrepreneurs and fast-buck artists saw an opportunity.

By the time Florida hit in 2003, gaming Google (and other engines) had become the full-time occupation of many an opportunist. And while some practices were perfectly legit—after all, what are publishers but content-based intermediaries between a customer and an advertiser—many were not. In the parlance of the Wild West, SEOs who took extraordinary and unsavory measures to game search engines became known as black hats.

At the same time, legitimate SEO businesses were also booming, with the goal of helping honest folks redesign their Web sites so that search engines could find, index, and accurately rank them. To aid them, Google and other engines published Webmaster guidelines outlining best practices for these optimizers. In short, the guidelines say "avoid black-hat practices."[2] Webmasters and business owners who followed these practices came to be known as white hats.

For the white hats, SEO was an essential part of doing business—after all, you want to make sure you put your best digital foot forward when it comes to search engines, and paying an SEO firm a thousand or two to ensure such a goal was a minor price to pay.

The trouble, of course, is that the early SEO industry was not entirely sure which practices were white hat and which were black. In fact, thanks to the rather vaguely worded guidelines on Google's site, coupled with the fact that the company keeps its algorithms closely guarded, SEO firms were increasingly tempted to push the limits of what *might* be considered white-hat practice. Many firms also made claims that were simply unreasonable—"pay me and I'll guarantee you are listed first in all major search engines," for example. The result: white-hat business owners unknowingly engage in black-hat practices, and their sites get banned from the Google index.

As discussed in previous chapters, the pre-Google search world also had no shortage of opportunists who took advantage of a search engine's ability to direct well-intentioned traffic to otherwise irrelevant sites—porn sites being perhaps the most visible offenders. But as search algorithms became more sophisticated, spammers had to adapt. PageRank rewarded sites with high-ranking inbound links and relevant anchor text, so spammers began to create link farms and doorway pages—essentially pages that did nothing more than link to other pages—so as to trick Google's index into assigning their pages (or in many cases, their clients' pages) a higher ranking for lucrative keyword search terms.

Google retaliated with ever more sophisticated algorithms, and the spammers counterstruck, blow for blow. Google banned certain IP addresses, for example, and spammers simply set up new ones.

But between white and black hats there is a significant area of gray, and it is in this maddening and capricious world that affiliate sites exist. Affiliate sites redirect potential customers to larger sites that have programs that pay for leads. Amazon and eBay, for example, have two of the largest and most profitable affiliate programs on the Internet. When a customer from an affiliate "converts" on the target site (buys a book on Amazon or an item on eBay), the affiliate gets a small cut of the action, usually no more than a few bucks. But far more profitable programs can be found from peddlers of prescription drugs and pornography, who will pay more than $40 for a new customer.

While no one would accuse eBay of being into pornography, in late 2003, the ecosystem that had sprung up around this Internet heavyweight smelled a lot like what might be termed black-hat spam.

Here's how. Most white-hat eBay affiliates pass along customers the old-fashioned way—from their own sites. For the purposes of this example, let's say that Mr. White Hat is a small merchant of carnival glass, and has a site devoted to this highly sought-after collectible. On his site he lists his wares, attaching descriptions and appraisals of value. He also links to eBay as part of his affiliate program. Every so often, one of his readers will head to eBay from his site, and should that reader convert to an eBay customer, Mr. White Hat gets a few coppers tossed into his eBay affiliate account.

But a more enterprising affiliate, whom we'll call Mr. Black Hat, realizes that the most lucrative place to find likely eBay customers is at a search engine—and in particular, among customers who are typing in keywords that might relate in one way or another to a product for sale on eBay. Mr. Black Hat then sets up doorway pages full of carnival glass keywords—essentially, content-laden Web sites that fool Google's spiders into believing his pages rank highly for a particular keyword. Thus a search on Google for "carnival glass" will show Mr. Black Hat's doorway page as a top result, pushing aside poor Mr. White Hat's pokey little carnival glass site.

This isn't a theoretical example. In fact, it happened just as I have described it, but on a major scale, in the fall of 2003. *AuctionBytes,*[3] a small publication covering the auction world, discovered that an affiliate by the name of Ryle Goodrich had created literally hundreds of thousands of these doorway pages, and was siphoning off thousands of dollars in affiliate lucre for his work. Even more damning, Goodrich had the implicit approval of eBay: he was sending so much lucrative traffic to the auction giant that the company had made him a preferred affiliate and granted him the right to republish eBay's auction listings—the very kind of content Goodrich needed to entice Google's spiders to his doorway pages. The coup de grâce? When a user clicked on the doorway ads, Goodrich converted the click to an actual search result within eBay. Hence, when searchers typed "carnival glass" into Google and clicked the first organic result, they ended up on a search page within eBay for the very same result.

When the *AuctionBytes* story broke, Google quickly took action and banned Goodrich's sites from its index. EBay, most likely wary of the bad publicity more than anything else, also took action to clean up its affiliates' practices.

Not two weeks later, Florida hit. Was there a connection? Perhaps, but more likely, it was simply one more example of an ongoing battle. Says Louis Monier, who now works as director of research at eBay: "Some of our affiliates are a little bit aggressive, but it's a general problem on the Web: any time someone will gain from traffic, they'll try to abuse the search engines. It's a well-known arms race. It was true in 1997; it's even more true today. My only comment would be: best of luck to Google."

Who Decides Shades of Gray?

Monier raises an important point. While Goodrich's approach clearly violated Google's guidelines, affiliate spam remains a major problem to this day—it's just far more sophisticated and difficult to track. For example, head to Google and type in "New York hotel." As of late 2004, most of the top results are what might be called remarketers—companies that essentially are arbitraging your desire to know more about New York hotels into possibly selling you a hotel room. Prior to November 2003, the same search was cluttered with affiliates who practiced black-hat tactics. So which is right? The practice of aggregating demand and converting it to sales is far older than the Internet (think travel agents, who get paid when they book your room), but the question remains: are these affiliates really what you were looking for when you typed that query in?

The term "digital camera" underwent an even more dramatic shift. As of this writing, the top results on Google are all review-oriented sites. But back in November 2003, before Hurricane Florida, they were mostly commercial sites looking to sell you a camera. Google seems to have decided that when you type that term into its engines, you only want to see reviews. Unless, of course, you take a look at the ads lining the right side of the results page. There you'll see all the folks trying to sell you stuff, clearly penned into their AdWords boxes.

This apparent contradiction lies at the heart of Google's algorithm-tweaking decisions—what might be called their editorial judgment.

Clearly the company has not targeted legitimate travel affiliate networks, but it has targeted eBay spammers. With cameras, Google has decided that folks who attempt to optimize their way into selling you a camera will be penalized, and only editorially oriented review sites can ascend to the much-vaunted first positions.

This raises a question, one that Google refuses to directly address: how does it make these decisions? How do you draw the line between pure, organic listings and paid listings?

Google's on-the-record answer is nearly always a variant of GoogleGuy's original posting on WebmasterWorld back in November 2003: we are always looking for ways to improve the quality of our rankings and algorithms. Clearly black-hat practices do nothing to better relevance, but hotel affiliates and digital camera reviews inhabit a grayer area: when is commercial speech no longer acceptable to Google's organic results?

The AdWords Connection

In the end, engines like Google reserve the right to determine what they believe is the best approach to relevance, and they will tweak their algorithms to ensure the results they feel are most relevant come first. It was clear that by the Florida update, Google had decided that affiliate and SEO spam had reached unacceptable levels. But Neil Moncrief and others had additional suspicions as to why so many blatantly commercial results suddenly disappeared from Google's organic results.

After Florida, Moncrief tells me grimly, "I had to buy AdWords. They forced me to do it." Taken together with AdSense, its syndicated cousin, AdWords accounted for about 95 percent of Google's billion-plus in revenues. After Moncrief dropped to five-hundredth for "big feet" and related searches, he had no choice but to buy his way back up to the top of the list. Otherwise, he'd face extinction. He did so, but with very mixed results. "[The ads] didn't work that well," he complains. That's not surprising: people tend to click on organic results far more often than they click on ads.

Moncrief was not alone in voicing suspicion that Google's direct attack on commercial spam ended up benefiting Google's bottom line. And while it's impossible to determine whether this fact affected Google's decision-making process (Google plausibly claims it did not), it's clear that affiliate spam was a drag on the company's AdWords business. To claim otherwise is to be naive. After all, why buy AdWords if your site gets all the traffic it needs from organic listings? Whether that was the main reason for the Florida update is a matter of contention. But that it helped Google's business can't really be disputed.

As Gord Hotchkiss, an SEO consultant, wrote on an industry site: "I really don't believe that Google purposely implemented the filter to drive advertisers to AdWords, but that is certainly a likely side effect. . . . If Google was targeting anyone with Florida, it was affiliate sites. A number of forum posts indicated that Google was taking aim at SEO. I don't believe so. I think Google is trying to wipe out bad SEO and affiliate programs and unfortunately there are a number of innocent bystanders who got hit in the cross fire."

Whatever its intent, Florida wiped out Moncrief's business. All of his keyword terms, every single one, fell from the first page of ranking to at least the fiftieth. And as anyone who's used a search engine knows, no one goes to the fiftieth page of results. One week before Thanksgiving, before the holiday period when Moncrief and just about every other online retailer make more than 80 percent of their profit for the year, the orders stopped coming in.

Damn right Moncrief bought some AdWords.

In the end, however, ecosystems tend to self-correct. I called Moncrief eight months after Florida, and he told me his business had once again risen to the top of the Google rankings for "big feet." How did he do it? "We hung in there, cleaned up the site a bit, and waited patiently," he told me. "We worked our way back up."

What about the last holiday season: did he get back in time for that? "No," he replied. "We had about four horrible months, and we lost the holiday season, which was the biggest part of our sales."

He paused, reflecting on the experience that, at least until the next time Google decides to dance, is in the past.

"It was a tough Christmas for the family," he concluded. Hurricanes come, he seemed to be saying. What can you do but pick up the pieces and rebuild?

The Marketing Firmament Shifts

Moncrief's story is one of a small-business owner tossed about on the seas of what has become a very big business, but if that were all there was to the search economy, this would be a very short book. In fact, search plays a much larger role in the world of marketing and commerce. Moncrief's small business is one portion of a far larger narrative.

Back when Moncrief was starting his business, the very idea of advertising on the Web was beyond his grasp. Banner ads were all the rage, and inventory was scarce, driven to outlandish prices by the rush to portaldom described in Chapter 5. But as those advertisers rushed for the doors during the crash, tens of thousands of Neil Moncreifs began to use services like Overture and Google AdWords. The simple reason? Paid search ads worked.

And why did they work? Because paid search shifted the marketing model from one based on content attachment to one based on intent attachment. In what might be termed the Web 1.0 version of online publishing, advertising followed a traditional, offline approach, adopting models that, in the main, borrowed heavily from print and television. Marketing messages were attached to content, whether that content was an online publication like *HotWired* or a Web-based service like AOL or Yahoo.

But the paid search ads pioneered by Bill Gross followed an entirely different model, that of intent attachment—a model best exemplified by the yellow pages or classified section of local newspapers. Initially, the marketers most driven to this new approach were large innovators like Amazon and small-business owners like Moncrief. And as paid search matures and the model begins to evolve, we can see outlines of a much larger shift occurring in the marketing business, a shift that is still in its early stages.

"The last bastion of unaccountable spending in corporate America." That's what Google CEO Eric Schmidt called corporate marketing when we last spoke. Google, of course, specializes in marketing that is entirely accountable—you pay only when someone clicks on your ad. Compared with the unpredictable and untraceable value of a magazine ad or television spot, search looks pretty damn compelling. But at the end of the day, three lines of text sitting next to a set of results is a meager way to declare your brand or inform a consumer about your new products or services. Clearly, there is room for both kinds of advertising—intent based and content based. But what if the two were to merge?

Before you dismiss the idea as mere speculation, let me lay out a scenario in which such a beast exists for the medium of television. First, imagine that a majority of households have a digital video recorder (DVR) of one kind or another (such a situation is predicted to occur within five years, according to Forrester Research). Further, imagine that this DVR has a search history of everything you've watched and are planning to watch (this is already done by most DVRs). Further still, imagine that this history is—with your tacit approval—blended with an edited profile of your online searching habits, forging a marketing précis of your likes and dislikes, your wants and needs (doing this is a matter of a marketing deal between DVR providers and search engines). Perhaps you use Google Desktop Search, or A9, or Ask, or Yahoo—it matters little; all of them can create such a profile already.

Now, let's set this scenario in motion. Let's say you are a young father-to-be. It's nine P.M. and your wife has settled, uncomfortably, onto her favorite couch. Clearing her throat, she politely reminds you that you've been a bit distant lately, that you haven't done a hell of a lot to help her around the house. You cringe. She continues: she's eight months pregnant, for God's sake, and when are you going to get around to reading that copy of *What to Expect When You're Expecting* that she left none-too-subtly in your briefcase six months ago?

Now, you're in your den, avoiding dealing with the sheer terror of becoming a father by checking your e-mail for the tenth time in so many minutes, but a pang of guilt finally reaches your irresponsible heart. So you start searching the Web, trying to get smart quickly. You Google "pregnancy baby" and head to the first link, Babycenter.com, where you read up on the eighth month. You then find a link to an article that lists ten things you can do to be a better husband. The fourth suggestion reminds you to read the books your wife has purchased, so you head to Amazon and buy another copy of *What to Expect When You're Expecting,* as you left the one your wife gave you next to the Gideon Bible on your last business trip.

"I'll read it, I promise," you tell your wife, and then add, "I'm on Babycenter right now, in fact." Pleasantly startled, your wife springs off the sofa—well, lumbers, perhaps—and peers over your shoulder. In a flash of inspiration, you intuit that there might be something you could watch together on TV that relates to the whole parenting thing. "Let's see if there's anything on TV that might be good," you say.

You click over to your TiVo home page, which lets you manage your television service much as you manage your weblog reading—through a search-based interface. You search for "parent childbirth newborn" or something like that and find that there are five shows in the next week that focus on the course of pregnancy, three of which are on the Learning Channel. You tell TiVo to record them all, noting that the first one will be available to download tonight, in half an hour, no less.

In the background of your computer, as you jump from site to site and page to page, several marketing-related actions are occurring. A cookie set by your local cable company notes that you've visited several sites that trigger marketing potentials—Amazon.com, TiVo.com, and Babycenter.com, all sites that indicate significant intent to purchase products or services. You've also alerted the system that you intend to download five new programs, and the system takes note of content tags associated with those programs, cross-referencing them with your recent search history.

The cable cookie shares this information with a marketing application running in the background of your computer, perhaps as part of that Google Desktop Search (GDS) program you downloaded last year. Alerted by the marketing potential that your recent surfing has created, GDS instantly uploads new tags to Google's central advertising marketplace—a marketplace that looks and works an awful lot like AdWords right now (for a refresher on that, head to Chapter 6).

Up on Google's ad marketplace, millions of similar potentials are aggregated and presented to hundreds of thousands of advertisers for sale in a modified real-time auction. Most of those advertisers have preset their spending levels, demographic preferences, and most important, intent-based targeting profiles. In the time it takes for an average Google search to finish—less than a second—several advertisements have already been sold against each of the five programs you've selected.

Half an hour later, you and your wife turn on your television to catch the Learning Channel show. As it starts, a small box appears on the bottom of the screen, alerting you to several advertisements that have appeared in your feed. You know that should you decide to watch them, your local cable bill will be reduced by a buck or so (or alternatively, you've selected the programming option that gives you free cable, but requires that you review ads at preset intervals). No matter, that's not really the reason you might want to pause the show and check out the ads. Turns out you rather like watching them, as they are often extremely relevant to your wants and needs, not to mention informative, linked as they are to robust Web sites and interactive features. So you pause the show, hit the ADS button, and scan the commercials.

Only they're not just commercials; they're offers as well. The first is from Gerber for a free month's supply of formula. (Pass; you and your wife have agreed that breast-feeding is the way to go.) Next up is a Pampers ad offering a free box of diapers. (Sure, why not? You accept that one, clicking the box that allows the system to send your details to the Pampers marketing machine.) Then comes the killer ad: "Click here for $50 off a Peg Perego Stroller. Ships in 24 hours!" *Huh?* you think to yourself. That's the one your wife said was the Mercedes of strollers. Maybe you can afford one after all.

"Honey," you begin. "What do you think? Should we go for it?" Her eyes light up (you had said no to this exact request twice—*$300 for a fucking stroller?!!* were your exact words) and you click to accept the offer. Your wife snuggles into your side, pleased that for once, her husband actually gets it. You return to the program, and . . . *scene!*

Is such a scenario possible? While the details will inevitably vary, I honestly think this scenario is not only plausible; it's inevitable. And it is the infrastructure of paid search as we understand it today that will make it all happen.

Now, look at this from the advertisers' standpoint. For an advertiser like Peg Perego, such a scenario not only makes television advertising affordable; it turns the medium into a *new sales channel.* Instead of buying time on the Learning Channel on Mondays at eight P.M. (a content-attached purchase), Peg Perego will buy direct access to the intentions you have declared through a blend of your search history and your television watching habits. Once it is satisfied that you are a potentially high-value customer, it will then push advertising offers down the cable line to your DVR.

The beauty of this scenario lies in how it changes the economic model of marketing. First of all, Peg Perego has never been a television advertiser, because the medium has never lent itself to a high enough return on investment—the company relies mostly on word of mouth and distribution through a network of retail outlets for its sales. But because it can identify exactly who its customers might be, on the basis of intent, it can change its model completely, and view a marketing investment in television to be, well, not a marketing investment at all, but rather—this is worth stating again—a new sales channel.

This in turn means that tens of thousands of marketers who otherwise may never have considered television a viable medium will soon see it as such. In the near future, it's quite possible that researchers tracking advertising by medium will have to fold television revenues into interactive—they'll often be one and the same.

That is the magic of intent-based marketing—it shifts marketing dollars from the unknown to the knowable. As Tim Armstrong, VP of advertising at Google, puts it, "search turns a cost center into a profit center."

I asked Armstrong what he thinks marketing will look like in ten years. His answer: "If you can imagine ten years from now every major and small advertiser with a totally digitized marketing asset set, so everything they can market is digitized with attributes against that—and they have hundreds of inbound and outbound feeds, and hundreds of places that either accept those feeds or pull them in. So in the future I think marketers are going to be agnostic about where their offers end up; they're going to be driven by the ROI [return on investment]. And I think most of the publishers on the Web [and think of the Web as including television] and most of the major other players on the Web are going to be able to put offers in front of people at exactly the right time. I think a lot of people today think Google and Overture when they think of ROI advertising. I'd like to think that in ten years they'd think only about Google, but more likely there will be ad systems in the back and tools that track ROI and conversions across multiple platforms and media. Advertising will be mostly margin driven."

Think about that for a minute. The entire foundation of marketing—$100 billion industry driving, well, nearly every business on this planet—is shifting, slowly but surely, to a new model, one informed by the simple idea of people looking for things on a search engine. No wonder Jan Pedersen, chief scientist for Search & Marketplace at Yahoo, recently quipped: "We think of shopping as basically an application of search."

They're All Search Businesses

But it's not just in advertising that search will have tectonic implications. To see how search has already changed an industry, consider the music business.

Let's start with the mother of all disruptors: Napster. According to Hank Barry, CEO of the infamous peer-to-peer service during the height of its controversial success, "Napster was, at its core, simply a search engine for music." In other words, Napster put the power of finding and acquiring music into the hands of consumers, and the entire music industry was consequently upended. Music is now a half-billion-dollar online business, showing no signs of stopping.[4] And anyone who thinks television and movies aren't next simply isn't paying attention.

Or consider the media business, and more specifically, the news business. Because of search, news has become fragmented—people can find news on nearly any topic delivered to them as lines of search results, as opposed to carefully laid-out stories on the front pages of local newspapers. In fact, Google News, a computer-generated news service offered free by Google, is now one of the largest news sites on the Web. What does it mean when news is no longer a destination, but becomes, thanks to search, a commodity? How can news pay journalists if there's no newspaper, per se, to purchase, nor a place next to which content-based advertising might be attached? Where, in short, is the ROI for news?

As a member of this industry, I certainly have given this a bit of thought. One evening, as I was decompressing after a long conference with Jonathan Weber, my editorial partner at the *Industry Stan-*

dard, and Steve Ellis, who runs an innovative music company called Pump Audio, I came upon some answers. Talk turned to what constituted quality content in a journalistic sense. Steve, who is British, asked Jonathan and me if we thought the *Wall Street Journal* represented the paragon of American newspaper feature writing. And I thought, *Jesus, I haven't read that paper for months.* I pay for the online version, but given how my reading habits have shifted, thanks to the online world,[5] the *Journal* simply has not crossed my radar enough to register.

Jonathan and I agreed that the *Journal* pretty much defined the American standard of good page-one feature writing, and I copped to being "*Journal* blind" for the past few quarters. Talk then moved to *The Economist.* Goodness, it had been ages since I had read that magazine either. I used to subscribe to the paper version (same for the *WSJ*), and when I did, I signed up for a few e-mail newsletters as well. But for whatever reasons those came intermittently, and they were not very good. Why, I wondered, were these two august bastions of journalism falling off my reading list?

You may have already guessed. Because they are fearful of losing revenue as a result of search, both require paid subscriptions, and therefore, neither supports the kind of deep linking that drives news stories to the top of search results at Google and its brethren. In other words, both are very difficult to find if you get your daily dose of news, analysis, and opinion from the Internet. And as we all know, folks who read their news on newsprint ain't getting any younger.

But there's more going on than just age trends. Media usage on the Internet is driven by different presumptions. In a print world, people read their own paper, then talked about the news when they got to the office or coffee shop. With the Web, however, news is a conversation—fueled by blogs, e-mail, and the cut-and-paste culture. In short, even if I did read the *Journal* or *The Economist,* I wouldn't discuss it nearly as freely as I would a story on Yahoo or Google News, because my friends and coworkers wouldn't be able to read what I read. More and more, I find that if I can't share something (that is, can't point to something using e-mail or my own Web site), it's not worth my time.

How does the news industry "cross the chasm" and survive in a search-driven world? I don't have a silver bullet, unfortunately, but it starts by opening up its sites and realizing that in a post-Web world, the model for news is no longer site driven. Sites that wall themselves off are becoming irrelevant, not because the writing or analysis is necessarily flawed, but rather because their business model is. In today's ecosystem of news, the greatest sin is to cut oneself off from the conversation. Both *The Economist* and the *Journal* have done just that.[6]

So what is to be done? My suggestion is simple: take the plunge and allow deep linking—let others on the Web link to your stuff. (The *Journal,* to its credit, has begun a limited implementation of this idea.) Notice that I did not say abandon paid registration; in fact, I support it. Publishers can let the folks link to any story they post, but limit further consumption of their site to paid subscribers.

I'd be willing to wager that the benefit of allowing the world to point to you will more than make up for potential lost subscribers. First off, publishers that do not offer additional paid subscription benefits beyond the articles themselves are not paying attention to the needs of their communities. In any case, many folks will pay to subscribe to a site that is continually being pointed to by sources they respect—be they friends sending links via e-mail, blogs, or other news sites.

In fact, I'd predict that the landing pages from such links might be the most lucrative places a publisher can capture new subscribers. It's a massive opportunity to convert: the reader has come to your site on the recommendation of a trusted source (the person who has pointed him to the story). It's pretty certain that if you make the page inviting, and use it as an opportunity to sell the reader on the value of the rest of your site (as well as show him some insanely relevant advertising), that reader will eventually feel the *Journal* is worthy of his support.

Why? In short, if readers find themselves pointed to the *Journal* on a regular basis, they know that by subscribing to the *Journal,* they would be more in the know. After all, many blogs read and point to the *Journal,* the reader thinks, so perhaps I should read it, too. Before subscribing, the only time a reader might find out something in the *Journal* is if someone points to it (a far sight from where things stand today, by the way). But if he subscribes, he can get his own feeds, and be first to know something. And, in the end, isn't that what drives subscription sales?

In the end, I think allowing deep linking will *drive* subscription sales, rather than attenuate them. Editors should not be worried about whether their content can "bring people to our site"—that's simply not a realistic approach anymore. The goal is to make content that is *worth pointing to.* If you're feeding the conversation, the rest will then follow, including advertisers who want to be in the conversation that news stories are fostering.

Local Is a Search Business

There is probably no greater example of a thriving off-line search business than the yellow pages. Standing at around $15 billion in the United States alone, the yellow pages are the print world's intent-based poster child. If you need a plumber (and don't already have a good one), where do you go? Well, if you're like most (older) folks, you pull out the yellow pages. Restaurant? Dentist? CPA? Dry cleaner? More than a billion times a year, Americans turn to their local yellow pages for the answer.

Within one generation, however, the yellow pages will be viewed as a dead industry.

Now, before you tell me that flipping though a printed directory is far more convenient than turning on your computer and punching in some search terms, let me remind you that local search, as it's called in the search industry, is still in its very early stages, and that the platform for local search—the PC-based Web—will not be the only, or even the primary, platform for this particular search-driven revolution. There will be about 1.7 billion mobile phone handsets in use by the year 2006, and most of those will have Internet access. When finding a dentist is as easy as punching "dentist" into your phone (or, with new technology already on the market, simply speaking it), the idea that anyone will pull out a ten-pound paperweight to execute a search will seem as quaint as hand cranking a car. When it comes to this market, local is entirely search driven.

The same goes for the classified advertising marketplace, which also stands at around $15 billion in the United States. Consider this fact: craigslist, a San Francisco–based company with fewer than fifteen employees, is now one of the top twenty Web sites measured by traffic. What does craigslist do? It offers classified advertising by local markets, currently more than fifteen of them. And what does it cost to post an ad on craigslist? Unless you are posting a new job, the answer is exactly . . . zero.

The mobile and local search revolution will have even more far-reaching implications for local retail shopping, however.

To explain, I'll need once again to sketch out a scenario, this one involving several elements already in place—search technologies, mobile phones, and the Universal Product Code (UPC) system—and a few more fanciful, but nevertheless feasible, technological and business model innovations.

Imagine it's the near future, and you're in your local grocery store on a mission to pick up dinner for a Saturday night dinner party. Because you've got oodles of disposable income to burn, it's a high-end Whole Foods store, the aisles dripping organic righteousness and whole-grain goodness. You know that dinner for eight is going to run you at least $200, not counting the wine, but that's OK compared with the tab at the local bistro. You'll be coming out ahead. But you do want to make sure you're not spending money you don't have to, especially on the wine.

Now, Whole Foods has quite a wine selection, but the store isn't known for its discount prices on anything, and when it comes to wine, you've got a sneaking suspicion that the store is really sticking it to you. But it's a convenience buy, you've always thought, and you're willing to put up with it for the most part.

As you slip your Neiman Ranch tri-tip into your basket and thank the butcher, you head to the wine aisle. What might go with that grilled tri-tip? A nice cabernet, no doubt. Whole Foods' wine aisle, a testament to hierarchy and peer pressure, places the most expensive bottles on the top, and the cheap juice on the bottom. No self-respecting Whole Foods shopper wants to be seen bending down to check out a bottle of wine. Then again, those bottles staring out at you from eye level are exactly the kind that you suspect Whole Foods marks up with the glee of a four-star sommelier.

What to do? Not to worry; you've got Google Mobile Shop installed on your phone. You whip out your Treo 950, the one with the infrared bar code reader installed, and you wand it over that $52 bottle of 2001 Clos du Val now lovingly cradled in your arm. In less than a second, a set of options is presented on the phone's screen. It reads:

> **2001 Clos du Val Merlot, Lot 21**
> **Stags Leap District, Napa Valley**
> Average Retail Price: $38 (click here for more)
> Click **here** for a list of prices at nearby stores
> Click **here** for stores selling similar items
> Click **here** for reviews of 2001 Clos du Val Merlot
> Click **here** for more on this vendor [ecological impact, vendor labor policies]

You're pretty sure that Clos du Val isn't employing child laborers, and anyway you're really interested only in price comparisons, and the first screen has confirmed your initial suspicion: Whole Foods is ripping you off.

You click on "list of prices at nearby stores" and see that the liquor store up the street is selling the same bottle for $39. You click on that store's link, and then choose the "reserve this item for same-day pickup" option. With a satisfied smirk, you replace the bottle on its perch on the top shelf, and head over to compare prices and recipe tips for $6 boxes of imported pasta. As you leave, the fellow who runs the store's wine department eyes you warily, then picks up the phone to talk to his manager. "Herb," he says. "Did you get my message about banning cell phones in the store?"

Is this scenario possible? For it to happen, a few nontrivial things need to occur. First, the entire UPC system must be made open and available as a Web service—a nontrivial event, to be sure. Those bar codes and the information within them are not yet a public resource, though a small coterie of researchers and entrepreneurs is looking to change that. Second, merchants must be compelled to make their inventory open and available to Web services. Third, mobile device makers must install readers in their phones, essentially turning phones into magic gateways between the physical world and the virtual world of Web-based information. And fourth, providers like Google must create applications that tie it all together.

While the first few hurdles to the realization of this scenario have yet to be jumped, it's certainly a no-brainer that Google and Yahoo would love to tie everything together should it become possible to do so. The implications of search breaking out of the PC box and making real-time information available at the point of purchase has been the failed business model of several Web 1.0 companies. But with recent developments in local and mobile search, it is far closer to happening now.

What might be the effects of such a system coming to fruition? For one, markets would have to compete far more on service, convenience, ambience, and other factors unrelated to price. And vendors of products that have been made in third-world sweatshops or in factories that overpollute, or vendors that support causes some consumers do not wish to support, would be called out in a far more transparent fashion. Refusal to participate in such a system would mean that vendors or merchants had something to hide, and so the system could be a major force for good in the global economy, forcing transparency and accountability into a system that has habitually hidden the process of how products are made, transported, marketed, and sold from the consumer.

It's All Search

Back stateside, search may not be a matter of freedom or control, but when it comes to the American economy, it is certainly changing industries. "Search has ruined the real-estate business," laments Martin Shore, a Marin, California–based real-estate developer. But Shore has a merry glint in his eye. He made his money in the pre-search days, back when in order to make a project work, you had to do the legwork yourself—get on a plane, scope out a neighborhood, talk to the people you might build a project for, and then lay your bets where your gut told you to go.

Now, however, reams of crucial information—title reports, demographic breakdowns, financial metrics—can be found on the Internet via relatively unsophisticated searches. As a consequence, the real-estate business has become far more competitive. "We used to go to places where young people and renters would hang out and ask them, 'Where would you want to live?' " Shore says. "Then we'd go to those up-and-coming neighborhoods and research the buildings—who owned them, how much were they going for, what was the title history. Finally we'd track down the owners and make them an offer based on a financial workup we had done. All that took a lot of work—it required relationships with the title company, with the people on the street."

But now, Shore says, information has replaced relationships. Because the information is easily available, the barrier to entering the real-estate business has been lowered, and thousands of new competitors have emerged—particularly during the recent real-estate boom. "People can sit in their offices in New York and find out just about anything they need to know about investing in Austin, Texas," Shore says. "You can't corner a deal by going to the physical location. Information travels faster than people on the ground. Deals are done sight unseen, based on information that is available to anyone."

The same could be said of nearly every information-mediated industry in the developed world—from travel to retail, banking to entertainment. Search has become the new interface of commerce.[7]

The Problems Looming

But not all is rosy in the search economy. Because of its innovative and relatively new business model, search is testing the boundaries of how business works in several ways. Nowhere is this more evident than in the field of trademark law.

Consider the case of a company by the name of American Blinds and Wallpaper Factory. This home-decorating specialist has built a $100-million-plus business in window coverings, wallpaper, and the like. As the search economy boomed, American Blinds profited from the rich stream of leads driven to its business from Google and other search engines. It quickly adapted its business model and recast its Web site as a one-stop shop for potential customers interested in redecorating their homes. It even trademarked its Web site's name: AmericanBlindsAndWallPapers.com, as well as the more memorable decoratetoday.com. In addition, the company began purchasing AdWords for generic terms like "blinds" as well as those based on the company's trademarked name: "American Blinds."

But in early 2003, American Blinds realized that while it owned the trademark on "American Blinds," it didn't own the market for it on Google's AdWords service. Competitors were snatching up the company's trademarks as AdWord terms (they did so by paying more for them, essentially), so that when customers typed "American blinds" into Google, they'd get advertisements for companies like JustBlinds.com and Select Blinds.

The company contacted Google and attempted to get the search engine to ban competitors from buying American Blinds' trademarked terms. Initially Google agreed to stop the practice, at least for terms that were, in fact, trademarked. But it refused to do so for what it deemed to be more generic terms, including "American blinds."

The trademark issue is far larger than just one company, and already sports a significant case history. Way back in the late 1990s, Playboy, Inc., sued Netscape for what was essentially the same kind of infringement. When searchers came to netscape.com and typed in "playboy," they would see banner advertisements from companies other than Playboy. The suit was initially dismissed, but later affirmed to go forward on appeal. Once it became clear that the suit would go to court (in early 2004), Netscape quickly settled. It knew it had a good chance of losing at trial.

Around the same time, Google filed its own complaint, asking a U.S. district court to, in essence, declare its AdWords policies legal. This was a reasonable attempt to forestall what the company could see would be a raft of lawsuits engendered by the Playboy ruling.

Google was correct to assume that lawsuits were on their way, and American Blinds was first in line. The company sued Google in early 2004. In May of the same year, Geico, a major insurance conglomerate owned by Warren Buffett,[8] also sued, along largely the same lines as American Blinds. Both cases are pending—the court has ruled partially in favor of Google in one instance, but the company faces more stringent standards in similar overseas cases. But no matter how they ultimately turn out, these cases represent a major cloud across Google's business model, as well as providing significant insight into the way that Google does business.

Trademark law is clear on what constitutes an infringement: any use of a competitor's brand to confuse or mislead a customer is verboten. In its suit, American Blinds claimed that Google was both encouraging and profiting from an illegal practice. Google countered that it was merely an intermediary, and that it could not be held liable for the actions of others. Google, the company argued, was simply a set of computer algorithms that worked without bias in the background.

Of course, Google *had* been selectively blocking trademarked terms prior to being sued, on a case-by-case basis (recall that Google did offer to ban the purchase of some of American Blinds' keywords). But in April 2004, the company issued a policy clarification, stating that it would now sell *any* trademarked term, no matter what. While the official reason given for the change was "better results," this clarification was clearly a legal gambit. If the company is to pursue a "we're just an intermediary" line in defending the trademark suits, it certainly could not be seen to be selectively protecting some trademarks, but not others.

This is where observers of Google's corporate culture get to see the company's Don't Be Evil motto put to the test. Google's PR machine whipped up spin that, to most dedicated observers of the law, was disingenuous at best. "By letting people restrict certain words, you're not getting the results that people expect from Google," Google VP Sheryl Sandberg told CNET News.com. In other words, this change had nothing to do with lawsuits, but rather was part of Google's ongoing mission to "better our search results." Limiting the sale of trademarked terms was tantamount to limiting free speech—that was the implication.

As one might expect, American Blinds' lead attorney David Rammelt sees it a bit differently. "If Google attempts to drape themselves in the flag of free speech, we'll be happy to show plenty of examples where Google was more than happy to limit speech if it was in their economic interest," he tells me.

Rammelt pointed to the case of Oceana, an environmental organization that purchased the keywords "cruise ships" and then displayed ads that directed consumers to a Web site eviscerating the cruise industry for anti-environmental practices. Google banned Oceana from purchasing those ads, citing a long-standing policy of not allowing "advocacy" advertisements (Google has since clarified, but not abandoned, this policy). But what constitutes advocacy is gray at best, and in any case, the practice of using commercial speech in opposition is rich and deep—as the pages of the *New York Times* on most given days will illustrate. Google, like any business, has the right to use editorial discretion over how and with whom it does business. But the fact is, Google finds itself held to a higher standard than most, because, as Neil Moncrief found out, Google is more than just another company. As far as the Internet ecosystem is concerned, Google is the weather.

Google is wading into a morass with its selective enforcements, Rammelt argues. And in the case of Oceana, it doesn't help the company's image that the travel industry is one of Google's largest advertising clients. In the end, Google's credibility comes down to one word: trust.

A Matter of Trust

These cases may test Google's ability to live up to its much-vaunted motto. Much is at stake. First is Google's—and by extension many others'—basic business model. It's difficult to estimate how large an impact an adverse ruling might have on Google's revenues, but it's fair to say it would be significant. Trademarked terms are the verbs of commercial speech.

But second, and perhaps more damaging, is what might come out during a protracted trial between Google and a well-funded adversary who has very little to lose and a lot to gain. "If we lose this case, we end up where we started," Rammelt told me. But if its adversaries win, Google will end up in the position of policing every trademark in the world, and losing an untold amount of revenue in the process.

Certainly that's enough to get Google's defenses armed. But the company stands to lose much more than that. Should cases like Geico and American Blinds go to trial, lawyers on the plaintiffs' side will dig up every arguable example of unfair and inconsistent behavior on Google's part, parading the evidence in front of what will certainly be a captivated international press corps. In short, these cases may well prove to be Google's equivalent of Microsoft's famed trial with the U.S. Department of Justice: a sapping PR nightmare that forever sullies the company's image.

And while Google may parry each example, such as the Oceana story, with its own spin and counterargument, there is one incident that may prove more troubling. If true, this story shows that Google, to further its own commercial interests, is willing to monkey with the one thing it said it would never compromise: the results it shows to consumers.

September 17, 2004, was the day the San Jose District Court was to hear arguments in the American Blinds case. This was not the start of the trial; far from it: Google had filed a motion to dismiss American Blinds' case (a motion which was later denied), and the judge had called both legal teams to his bench to argue their positions on the motion. This was the only chance both sides had to convince the judge of the validity of their arguments.

The morning before the arguments, a member of American Blinds' legal team sat alone in his hotel room, fiddling with his computer, trying to get the hotel's broadband to work. To test the system, he brought up Google and entered what had become a habitual search query: "American Blinds." After all, that was the whole reason he was in this sterile hotel room, 1,500 miles from home: every time someone entered "American Blinds" into Google's search field, competitors to American Blinds came up on the screen.

Only this morning, for some reason, they did not.

That morning, the results for "American Blinds" on Google were entirely innocuous. The only paid sponsored link for the query was American Blinds' own advertisement. The lawyer was stunned. He checked again and again. Nothing but good, clean search results, with nary a potentially trademark-damaging result in the bunch.

The lawyer suspected Google had changed its results, and called colleagues in other parts of the country, who repeated the "American Blinds" search. Sure enough, searches in other regions returned different results, including the potentially infringing advertisements. The lawyer couldn't believe it: was Google intentionally sanitizing results in the San Jose region so as to sway the court's opinion on the case? And was the company so arrogant that it thought it could get away with it?

The lawyer quickly documented his findings, instructing staff members to take screen shots proving that the search results were different in San Jose from the rest of the country. If ever there was a smoking gun, he thought, this was it.

The next day in court, Google's and American Blinds' teams argued pro and con on the motion to dismiss. Toward the end of the hearing, the American Blinds lawyer dropped his bomb: he had what he believed to be incontrovertible proof that Google had fiddled with its own search results *this very day* and *only in this region* so as to sway the court's opinion in this matter. "Jaws dropped over on Google's legal team," the lawyer recounted. "Trust represents the keys to Google's kingdom. Google works only if its customers believe it is unbiased and fair."

To be clear, this kind of fiddling is absolutely sacrilegious at Google, and the company has made repeated statements along those lines, to me and anyone else who might bother to ask. When I asked Google PR for a response to the American Blinds allegation, a Google spokesperson told me that "Google would certainly never do such a thing." How then might he explain the lawyer's allegation? The spokesperson told me he did not know; perhaps it was a technical glitch.

Others familiar with the allegation question why Google would engage in what would clearly be damaging behavior should the company be caught. After all, the company is claiming that running competitive ads based on trademarked terms *should* be legal. I asked the lawyer to respond to that reasoning. "I suppose a cynic might think that it was done to lessen the visual starkness of the confusion that results when all these competing Web sites pop up after you type in 'American Blinds,'" he said. "Ultimately, to prevail in our case we have to prove there is a likelihood of confusion. The judge that day probably would not see a lot of confusion if he tried that particular search."

The judge declared that this new allegation was not pertinent to the motion hearing at hand, as it was based on alleged facts, and should therefore come out during discovery, a phase of the trial set to begin in late spring 2005.

Should the trial go forward, the allegation related above will hit every newspaper, Web site, and telecast in the free world. Is that enough to sink Google? Certainly not. But just ask Microsoft—and its shareholders—what effect the *U.S. v. Microsoft* trial had on the once high-flying company. The answer can be found in the company's stock price, which hasn't risen since the case was filed nearly five years ago.

But it is far more likely that this allegation of Google's index fiddling will remain just that, an allegation, unsubstantiated by the credibility of a court ruling or any specific evidence that Google purposely manipulated its index. Depending on how the case proceeds (there are several similar cases pending), Google can always modify its policies regarding trademarked terms and settle the American Blinds case. In the end, it's fair to say that however the trademark issue is resolved, the search economy will continue its breakneck growth and ongoing conquest of new commercial terrain. Unless, of course, click fraud doesn't stop it cold.

Click Fraud

It's fair to say that click fraud threatens to undermine the entire premise of Google's and Yahoo's success. Click fraud is the decidedly black-hat practice of gaming not organic results (as in the case of the eBay affiliates), but paid search ads, the very heart of the search economy.

In short, purveyors of click fraud take advantage of the syndicated nature of Google's, Yahoo's, and other search providers' advertising networks. They sign up as Google AdSense publishers, for example, which allows them to distribute Google's advertisements alongside their own content. But instead of running real content, these black hats run only AdSense ads on their sites. They then run robots (or low-wage workers in India or Eastern Europe) over those pages, mechanically clicking on every single ad, earning a cut for themselves and a cut for Google. The unwary advertiser pays the freight.

Click fraud is as old as paid search; in the course of reporting this book I spoke to people who recall the problem plaguing GoTo.com back in the late 1990s. Most search engines could deal with it as it came up—as soon as they found a fraudulent publisher, they'd shut down its account. But because Google's AdSense has such wide distribution—distributed as it is to hundreds of thousands of publishers—it's nearly impossible for the company to stay ahead of new scams. Many advertisers claim that up to 25 to 30 percent of their budgets is lost to click fraud—a figure that Google does not dispute, but calls an "outlier." "The average [amount of click fraud] is far lower than than that," says Salar Kamangar, who runs Google's advertising programs. He points out that like Yahoo, Google employs a wide array of anti–click fraud tools, ranging from algorithms that discover fraudulent sites to teams of humans who follow up on advertiser complaints.[9]

Some level of click fraud is to be expected—one can reasonably expect that an irate customer or competitor may want to hurt a business by repeatedly clicking on its paid links, thereby exposing the victim to unexpectedly high marketing bills.

But intentional, robot-aided click fraud is a far more virulent strain of cheating, and despite Google's and Yahoo's best efforts to contain it, at the time of this writing, it represented a mounting threat to both companies' core business model. "Something has to be done about this really, really quickly, because I think, potentially, it threatens our business model," Google chief financial officer George Reyes told an investor conference in December 2004. "There's a lot of bad guys out there that are trying to take advantage of this."

For a sense of how bad it might get, look no further than the second-tier players in the search market—the Mamma.coms and Findwhats of the world. According to one former executive at another second-tier search network, more than 40 percent of his engine's clicks were most likely click fraud. "That's forty percent of my company's revenue," the executive told me. When the executive asked his CFO what the company was going to do about it, he was told to keep the matter quiet. No company can afford to lose 40 percent of its revenue, after all.

And therein lies the rub of click fraud. Every time someone clicks on a paid search ad, the search engine gets paid. From a short-term financial point of view, a little click fraud is good for business. But in the long term, it benefits no one to allow fraud to flourish. Bribery, payoffs, and fraud are rampant in the early stages of nearly every emerging capitalist economy—from the Wild West to modern-day Russia. The search economy is no different. But eventually, the rule of law prevails. Among the first-tier companies—Google, Yahoo, Microsoft—search fraud is already taken extremely seriously, and efforts to combat it are intensifying. "We'll never turn a blind eye to this," says Patrick Giordani, who runs loss prevention at Yahoo's Overture subsidiary. "Our goal is to stop it all."

Chapter 8
Search, Privacy, Government, and Evil

This will go on your permanent record.

—The elementary school principal

Did you know that Google knows where you live? Worse yet, did you know that Google *will give out your address to anyone who asks*? Who the hell does it think it is?

Given that I write about search, a fair number of alarmist e-mail threads are forwarded my way—some by friends, others by colleagues, but often with the same revelation: Google knows where you live. By the time they've gotten to me, the e-mails have wound their way fairly well through the six-degrees-of-separation Web, CCed and forwarded to scores, if not hundreds, of souls. The subject line usually blares something along the lines of "I can't believe they can do this!" or "Oh my God, did you know?"

Here's a sample e-mail, with identifying information deleted:

> Subject: This is hard to believe, but true, I tried it.
>
> Google has implemented a new feature wherein you can type someone's telephone number into the search bar and hit enter and then you will be given a map to their house. Before forwarding this, I tested it by typing

> my telephone number in google.com. My phone number came up, and when I clicked on the MapQuest link, it actually mapped out where I live. Quite scary.
>
> Think about it—if a child, single person, ANYONE gives out his/her phone number, someone can actually now look it up to find out where he/she lives. The safety issues are obvious, and alarming. This is not a hoax; MapQuest will put a star on your house on your street.

It's easy to understand the initial reaction this feature elicits. You type in your phone number—a uniquely personal form of identification—and up pops a map of your house. First reaction for those who've never seen such a thing before: *my God, they know where I live!* And this fear of such a simple thing—known as a reverse directory lookup—bears further contemplation.

In our society, reverse directories are legal. Addresses and phone numbers are presumed to be public information, unless the resident requests an unlisted number. As much as we might like it, we can't make our physical address private, though there are certainly other ways to avoid tying your personal identity to where you live, should you wish to. Connecting a phone number with an address is also legal—reporters, cops, and private detectives do it all the time.

But while this kind of information is public, it is not widely available. Until Google and others made the digital connection via search, the public could assume it was difficult to do a reverse directory lookup, and only those with explicit or tacit societal permission—law enforcement or the fourth estate—ever took the time to do so.

American society was built on the enlightened and somewhat thrilling idea of the public's right to know. Our government is meant to operate more or less in the open. The same is true of our courts: unless a judge determines otherwise, every divorce, murder, felony, misdemeanor, and parking ticket is open to public scrutiny.

But while it's comforting to know that we, the public, have the right to review this information, it's also comforting to know that we very rarely do. After all, regardless of your prurient desire to know whether your new coworker has a messy divorce or a DUI in his otherwise well-appointed closet, most of us will not waste an afternoon down in the basement of our county courthouse to find out. The very fact that it's so much trouble to find such information has, in effect, muted that information. Unless office gossip precedes our new partner in cubicle land, we don't even think of such questions when introduced to our new coworkers.

But what if it were as easy as typing his name into Google? Often, it already is. If your cubicle mate happened to have a messy divorce, one covered in the papers or simply added to digitally available civil case files (many jurisdictions do just that), it won't be very hard to find. Or perhaps he spurned an ex-lover with a blog and a grudge, a lover who has turned their spat into a permanent entry in the Database of Intentions. Or maybe your office mate was slapped on the wrist by a professional organization, a rebuke noted in that organization's monthly newsletter, which now lives online.

Such is the case of Mark Maughan, a Los Angeles CPA who Googled himself and didn't like what he saw. His vanity search listed a page from the California Board of Accountancy noting he had been disciplined professionally, a claim he disputes. Maughan has sued Google, Yahoo, and various other search engines, though his suit is widely expected to fail (as to why, in short: don't blame the messenger). The lesson, however, is clear: in the minds of others, you are what the index says you are. If you don't like it—well, change the index. Oddly enough, all the coverage of Maughan's suit has done just that—pushing the offending page lower, but raising Maughan's controversial profile even higher. The first relevant result now for "Mark Maughan" on Google is a blog post from a site called Overlawyered that excoriates Maughan for filing what the site believes is a frivolous case.

The examples of this public privacy issue go on and on. As anyone who has lost or found a loved one knows, there is no more powerful search than a search for a person. Take the case of Orey Steinmann, a seventeen-year-old who typed his name into Google and discovered that his mother—with whom he was still living—had abducted him when he was a toddler. Turns out his mother had lost him in a custody battle, so she fled from their home in Canada to Southern California, where mother and son lived without incident until Steinmann vanity-Googled himself and learned that his father had been looking for him for nearly fifteen years. After that fateful search, Steinmann told his schoolteacher, who told authorities. His mother ended up in jail, and he has not spoken to her since.

Of course, search can turn up some titillating stuff as well—like the case of the ugly divorce in San Diego, California. According to an August 2004 article in *Forbes,* a couple in the midst of a nasty divorce discovered that the details of their rather rancorous proceedings—including the husband's income, the wife's predilection for furs, and the husband's desire to marry again—were up for all to see on Google (the information has since been taken down).

The simple fact is this: nearly everyone with access to a computer will Google someone else. If you are a knowledge worker, chances are you Google someone nearly every day, if not more often. Have a job interview? Google the prospect. Want to get ahead with your boss? Google her before your next review. Got a date with someone new? Google him—you never know if he might be wanted by the FBI. A woman in New York City did just that to LaShawn Pettus-Brown, a man she was to meet for a first date in a restaurant. When she saw that the man was wanted by the FBI, she alerted authorities, who met the man and arrested him.

Given the ubiquity of search, soon everyone will be Googling everyone else. What might it mean if someone *isn't* in the index? Does that mean he is of a certain class, either too low to be noticed by search's insatiable spiders, or so rich as to be able to avoid them altogether? Certainly such a person—a person who is not in the index—will have a certain air of mystery before too long.

For the rest of us, it's a good idea to check your own name on Google, early and often. Given that just about everyone else you meet will be doing it anyway, it's just smart to get a picture of who you are in the world according to the index. In the Google age, every new relationship begins with a Google search.

What do we do when information that we know, by law, should be public, becomes, well . . . *really* public? As in first-page-of-links-when-you're-Googled public? What happens when every single thing that's ever been publicly known about you—from a mention in your second-grade newsletter (now online, of course) to the vengeful ravings of a spurned lover—trails your name forever? Should we as a society legislate away the digital, and draw the line of what's public at information on paper, stored in a musty clerk's office?

In fact, the Florida Supreme Court considered that very question in late 2003, and came down on the side of caution—limiting electronic access to public records pending a full review due sometime in 2005. Clearly, this is an unresolved issue in our society.

As digital information spreads and is connected through search, unexpected challenges arise, challenges that conflict with presumptive and rarely voiced social norms. The reverse directory lookup illustrates a particularly discomforting expression of this public privacy issue. Search engines like Google both create and expose this issue, reminding us of conflicts between the law and the mores to which we've become accustomed. We're fine with folks knowing our phone number—we know it's pretty much public record. But the act of using technology to connect that number to our address, our home, the place we keep most sacred—that's somehow out of bounds. Thanks to search, we must confront one of the most significant and difficult issues a democracy can face: the balance between a citizen's right to privacy and someone's—be it a corporation, a government, or another citizen—right to know.

Or, as many privacy advocates fear, perhaps it has nothing to do with a *right* to know—but rather simply the *ability* to know. In the 1967 science fiction classic *Chthon,* author Piers Anthony imagines a dictatorial future civilization where all knowledge is universally available via computer. Mainly for historical reasons, however, society has kept a massive storehouse of books—traditional library stacks. In an attempt to track down a mystery, the novel's protagonist decides to go to the stacks as opposed to querying the computer system. Why? He knows that if he uses the paper stacks, no one can trace his actions, and he won't alert the authorities.

The fact is, massive storehouses of personally identifiable information now exist. But our culture has yet to truly grasp the implications of all that information, much less protect itself from potential misuse.

Search Me

Google learned of this situation the hard way in mid-2004, when it introduced the beta version of Gmail, a new e-mail service that boasted 1 gigabyte (1,000 megabytes) of storage. Google fully expected the product to be a hit—after all, Web-based e-mail programs from Microsoft and Yahoo had measly 10-megabyte limits, and they charged if you wanted more. Gmail leveraged Google's core asset—its technology infrastructure—and completely rewrote the rules of the e-mail game. Not to mention that Gmail had a Google-like search interface that was arguably far better than its competitors'.

But instead of basking in the glow of adulatory press, Gmail sparked the company's first full-blown PR crisis. The reason? Privacy. Gmail used Google's AdWords technology to place advertisements alongside users' e-mail messages. Now, the idea of placing ads in e-mail is certainly not new—Yahoo and Microsoft both did it, and Web mail users were accustomed to seeing ads—they were the quid pro quo of having a free service. But somehow Gmail pushed the boundaries—its ads were simply *too relevant*. When Mom sends an e-mail about apple pie to her son, and her son sees ads for apple pie recipes alongside her e-mail—well, for some, that crosses the line into spooky. It's a transgression of the public-private line—it's as if someone at Google was really reading Mom's e-mail, then choosing the ads that should accompany it.

The initial reaction was negative. "Search is one category; your e-mail is quite another. Do you really want Google snooping so close to home?" wrote Charles Cooper, a commentator on CNET.com, an industry news site. "The company says it is not going to read the contents of anyone's in-box. Still, you don't need to be a privacy extremist to realize that this fundamentally remains a bad idea."

Of course, Google's computers were not actually *reading* the e-mail; instead, they were simply parsing text for matches with the AdWord network. And that's the difference between Google's approach and that of Yahoo or Microsoft: Google used e-mail as a distribution system for its massive network of advertisers. Since there were so many possible ads for any given phrase, the chance that one matching an otherwise innocuous line in e-mail ("apple pie," for example,) would come up was quite high, compared with the more primitive approach taken by other e-mail providers.

To most of the world, it appeared Google was indeed reading your e-mail. Now, to be clear, only human beings can actually read,[1] but that distinction was for the most part lost in the ensuing debate. And there were larger issues at play. Privacy advocates such as Daniel Brandt of GoogleWatch.org pointed out that now that Google had your e-mail address, it could potentially tie your IP address (a unique number that is used by browsers to identify your computer) to your identity, creating an opening for all sorts of potential privacy abuses. Theoretically anyway, Google could now track your entire Web usage, not just your e-mail.

Sensing an opportunity to make headlines, California state senator Liz Figueroa introduced legislation to ban Gmail outright. "Figueroa Introduces Bill to Stop Google from Secretly 'Oogling' Private E-Mails" read a press release announcing the bill.

The bill got plenty of press and sparked vigorous debate, and at this writing an amended version—no longer banning Gmail but rather adding protections from corporate snooping—has passed the California Senate and is pending a House vote. E-mail, we have all realized, is moving from the ephemeral to the eternal, becoming one more record in the Database of Intentions that might be indexed and served for all the world to see. Regardless of whether the California bill passes, Gmail hit a nerve—for the first time, people realized their very private thoughts are subject to the scrutiny of a technological infrastructure that was quite literally out of their control.

As if to drive home the reach of technology into everyday life, not six months later Google introduced Google Desktop Search, a program that indexes your entire hard drive much as Google indexes the Web itself. GDS, as it became known, was followed by desktop search products from every major search player, from Ask to Yahoo. While desktop search did not raise the same level of public handwringing as Gmail, the fact remains: once you index the contents of your computer using desktop search, your private information is far more accessible. In fact, GDS even goes so far as to make it appear that the contents of your desktop are integrated into its Web-based service. In fact, your data stays on your hard drive, but the technology to upload it to the Web is trivial. Only Google stands between your privacy and the will of a determined hacker or government agent.

But desktop search and Gmail are not the only examples of how our digital private lives might collide with the public realm. Internet service providers (ISPs) and universities (which act as ISPs for their students and staff) regularly keep records of where their users go, what they search for, and when they are using the Internet. Search engines keep voluminous logs of user interactions, mainly to divine patterns to make their engines more efficient and profitable. Will all these new records ever be indexed and made publicly available? Probably not. But what happens when they fall into the hands of the wrong people, or even those with good intent, but poor judgment?

And at its heart, privacy is about trust. By using Gmail, Google Desktop Search, Hotmail, or any other service that connects your computer and its contents to the Web, you no longer totally control how your private documents, your communications, or even your own browsing history might be used. Like it or not, you are now in a relationship of trust with your service provider. Sure, Google's motto is Don't Be Evil, and sure, all good organizations have privacy policies, but they vary widely and have exceptions that can be interpreted in any number of ways (and who really reads them, anyway?). All companies, for example, can be compelled to deliver information about you should they be presented with a court order. And many businesses reserve the right to review your personal information if they suspect you are acting in a manner contrary to their internal policies.

Do you trust the companies you interact with to never read your mail, or never to examine your clickstream without your permission? More to the point, do you trust them to never turn that information over to someone else who might want it—for example, the government? If your answer is yes (and certainly, given the trade-offs of not using the service at all, it's a reasonable answer), you owe it to yourself to at least read up on the USA PATRIOT Act, a federal law enacted in the wake of the 9/11 tragedy.

Unreasonable Search?

The USA PATRIOT Act[2] was introduced into Congress one week after the September 11 attacks, then signed into law not six weeks later—breathtakingly fast by Washington standards. The legislation amended nearly twenty federal statutes and lacked the typically moderating force of legislative debate—the PATRIOT Act was the Bush administration's first official response to September 11, and few were willing to be on record opposing it. After all, we were under attack; this was war; all bets were off.

But as calm returned to Washington and legislative watchdogs (and the press) began to chew through the act, some disturbing facts began to emerge. First, the PATRIOT Act was in many respects a rehash of the Anti-Terrorism Act of 2001 (ATA), an extremely controversial piece of legislation that had been stuck in draft form for months prior to the attacks. And for good reason: ATA significantly expanded the government's ability to access and monitor private information—the very kind of information found in your e-mail, in your search history, and on your Google Desktop Search application. While the Bush administration was eager to get ATA passed, there was simply no way it would, at least not without significant revisions and added protections. But when 9/11 hit, the Bush administration dusted off ATA, revised it, then resubmitted it as the PATRIOT Act.

So what exactly does the PATRIOT Act do? The act revises several previous privacy and government surveillance–related acts, extending federal authority to a number of new areas, including the Internet. It redefines several key terms in these prior acts—particularly those concerning phone-tapping devices called pen registers and traps—so as to broaden their scope. Bush administration officials argued that these revisions simply brought the law from the telephone age to the Internet age, but the truth is a bit more nuanced than that. According to an analysis from the (admittedly anti-PATRIOT) Electronic Privacy Information Center (EPIC):

Prior law relating to the use of such devices [pen registers and traps, which record telephonic information] was written to apply to the telephone industry, therefore the language of the statute referred only to the collection of "numbers dialed" on a "telephone line" and the "originating number" of a telephone call. The new legislation redefined a pen register as "a device or process which records or decodes dialing, routing, addressing, or signaling information transmitted by an instrument or facility from which a wire or electronic communication is transmitted." A trap and trace device is now "a device or process which captures the incoming electronic or other impulses which identify the originating number or other dialing, routing, addressing, and signaling information reasonably likely to identify the source or a wire or electronic communication."

By expanding the nature of the information that can be captured, the new law clearly expanded pen register capacities to the Internet, covering electronic mail, Web surfing, and all other forms of electronic communications.

In other words, under the PATRIOT Act, the government now has far broader rights to intercept your private data communications—a reinterpretation of the Fourth Amendment, which states: "The right of the people to be secure in their persons, houses, papers, and effects, against unreasonable searches and seizures, shall not be violated."

The PATRIOT Act certainly puts a new spin on the word "search." But this is to be expected, right? After all, if the government has probable cause and a search warrant, nothing has really changed, has it? As all good civics students know, the Fourth Amendment continues: "no warrants shall issue but upon probable cause, supported by oath or affirmation, and particularly describing the place to be searched, and the persons or things to be seized."

Under PATRIOT, prior interpretations of these constitutional presumptions don't necessarily hold true. To summarize, the PATRIOT Act holds that your private information can now be intercepted and handed over to government authorities not via a search warrant tendered to you, but rather via a request to your ISP, your community library, or another service provider. That means that should the government decide it wants access to your information, it no longer needs to serve a search warrant on you; it can instead go to the company that you use—be it Google, Yahoo, Microsoft, AOL, or any number of others.[3] In the past, the government could certainly tap your phone or search your effects if you were a suspect in a crime. But under the PATRIOT Act, not only can the government tap a suspect's clickstream; the standards for who the government can tap and how it informs a suspect have loosened as well.

OK, you might respond, that's all well and good, but certainly the government has to declare reasonable cause for searching my stuff, and if I'm not suspected of a crime, I will be notified, right? According to PATRIOT, the answers to those points are not really, for the first, and emphatically no for the second. PATRIOT specifically prohibits companies from disclosing to *anyone* that the government has requested information from that company, effectively drawing a curtain around our government's actions. And while PATRIOT does require that a court find "reasonable cause to believe that providing immediate notification of the execution of the warrant may have an adverse effect," and that the government eventually must inform you that you've been searched, the standard for what is reasonable cause or notice is not stated.

By now, you might be a bit concerned about abuse of power under the PATRIOT Act, but you're not a foreign agent bent on the destruction of the United States, and the law is really only interested in foreign agents, after all.[4] Most of this stuff doesn't apply to you, does it? In fact, PATRIOT changes the law so that government officials no longer have to prove they are after a foreign agent when they intercept communications. Now, all they have to prove is that they feel access to your information might be valuable to their investigation. That's a pretty broad stroke. Fortunately, a provision was added that prohibits surveillance "solely on the basis of activities protected by the First Amendment." But how does one tell the difference between your First Amendment right to do searches about the tactics of terrorists, for example, and the searches of a real terrorist?

That's a hard one.

One might argue that while the PATRIOT Act is scary, in times of war citizens must always be willing to balance civil liberties with national security. Most of us might be willing to agree to such a framework in a presearch world, but the implications of such broad government authority are chilling given the world in which we now live—a world where our every digital track, once lost in the blowing dust of a presearch world, can now be tagged, recorded, and held in the amber of a perpetual index.

Not surprisingly, a backlash has begun to build against the PATRIOT Act. In an extraordinary move, New York, the city most wounded by the 2001 attacks, passed a resolution repudiating the act. By doing so, the city council of New York joined nearly a dozen state and local government agencies that have passed similar measures. Since this resolution comes from a place wounded by the very acts of terror PATRIOT was reputedly passed to prevent, excerpts from the resolution's words bear consideration.

Whereas, The City of New York has a diverse population, including immigrants and students, whose contributions to the city are vital to its economy, culture and civic character; and

Whereas, The members of the Council of the City of New York believe that there is no inherent conflict between national security and the preservation of liberty—Americans can be both safe and free; and

Whereas, Government security measures that undermine fundamental rights do damage to the American institutions and values that the residents of the City of New York hold dear; and

Whereas, federal, state and local governments should protect the public from terrorist attacks, such as those that occurred on September 11, 2001, but should do so in a rational and deliberative fashion in order to ensure that security measures enhance the public safety without impairing constitutional rights or infringing on civil liberties. . . .

. . . Resolved, That the Council of the City of New York opposes requests by federal authorities that, if granted, would cause agencies of the City of New York to exercise powers or cooperate in the exercise of powers in apparent violation of any city ordinance or the laws or Constitution of this State or the United States; and be it further

Resolved, That the Council of the City of New York urges each of the City's public libraries to inform library patrons that Section 215 of the USA PATRIOT Act gives the government new authority to monitor book borrowing and Internet activities without patrons' knowledge or consent and that this law prohibits library staff from informing patrons if federal agents have requested patrons' library records . . .

The resolution goes on to demand that federal officials who make information requests under the PATRIOT Act's veil of secrecy be held accountable, and that citizens who have been investigated without their knowledge be informed.

Several lawsuits have been filed challenging the constitutionality of the PATRIOT Act, and the act will be up for renewal in the fall of 2005. Regardless of how or whether the act is renewed, its initial passage is certainly thought-provoking as we all enter the age of search.

In early 2005, I sat down with Sergey Brin and asked what he thinks of the PATRIOT Act, and whether Google has a stance on its implications. His response: "I have not read the PATRIOT Act." I explain the various issues at hand, and Brin listens carefully. "I think some of these concerns are overstated," he begins. "There has never been an incident that I am aware of where any search company, or Google for that matter, has somehow divulged information about a searcher." I remind him that had there been such a case, he would be legally required to answer in just this way. That stops him for a moment, as he realizes that his very answer, which I believe was in earnest, could be taken as evasive. If Google had indeed been required to give information over to the government, certainly he would not be able to tell either the suspect or an inquiring journalist. He then continues. "At the very least, [the government] ought to give you a sense of the nature of the request," he said. "But I don't view this as a realistic issue, personally. If it became a problem, we could change our policy on it."

But while the PATRIOT Act has significant implications for the government's ability to leverage corporate information for its own purposes, there are other concerns as well.

"There are multiple paths to hell," observes Lauren Weinstein, a longtime Internet privacy advocate and computer engineer. "We have tended as a society to think of the government as the entity that might build an Orwellian database. But the private sector might just do it, and in a far more powerful way."

According to Weinstein, we need not live in fear of an all-knowing Big Brother. Instead, we should live in fear of any entity that possesses the *ability to know* whatever it wishes to know, should the need ever arise. One such entity is ChoicePoint, a commerical data aggregation company that holds detailed records on hundreds of millions of people. ChoicePoint is just one of scores of similar companies. In early 2005, ChoicePoint became the subject of intense scrutiny when it was discovered that the company had sold personal information to identity thieves. Journalists were quick to point out that besides the fraudsters, one of the company's most reliable clients was the U.S. government.[5]

But another kind of data aggregator is your ISP, your mail provider, or possibly your search engine. According to sources Weinstein claims to have inside the company, Google regularly works in an informal fashion with law enforcement agencies, tracking down personally identifiable information for authorities without notification to the person involved. In addition, Weinstein claims that Google engineers often track personally identifiable information to test new products and services, as well as to simply "play"—to do pure research to test the limits of what is possible with the information at Google's disposal. As a policy, Google refuses to comment on its relationship with law enforcement or its use of its vast storehouses of data, but a spokesperson did point me to the company's privacy policy.

Google's privacy policy allows the company to review your personal information, should it decide it wishes to do so. From that policy:

We may share [private] information . . . [if] we conclude that we are required by law or have a good faith belief that access, preservation or disclosure of such information is reasonably necessary to protect the rights, property or safety of Google, its users or the public.

While Google's public image is that of a sunny company that will never do evil, this policy gives the company extraordinary latitude with regard to your personal information. It also lays the definition of "good faith" and "protection of the rights of the public" squarely with Google, rather than a court order or the government. In other words, if Google decides that tracking and acting upon your private information is in its best interest, it can, and it will.

While our government is—at the end of the day—accountable to the people that fund it and elect its leaders, a public company, even one as well-intentioned as Google, is accountable to two forces: its leaders and its shareholders. And at no company are policies immutable.[6]

The China Question

Then again, at least we don't live in China. In response to the perceived threat that search and the Internet represent, China has gone to extraordinary lengths to censor the Internet—to the point of building what is known in academic circles as the Great Firewall of China, a technological infrastructure that automatically filters out banned sites—political opposition sites in Taiwan or Tibet, for example—from the walled garden of the Chinese Internet.

Search companies have long had to deal with the laws of other nations—because of local regulations, Google and Yahoo filter Nazi hate sites from their local indexes in Germany and France, for example. But China takes a far more unbridled view of what it considers dangerous information.

"China is a curious hybrid, a miscegenation of Leninist institutions and political structures imported and established in the fifties during the Stalin era and a more recent importation of dynamic market structures and values," says Orville Schell, a China scholar and dean of the Graduate School of Journalism at the University of California, Berkeley. "There has been great economic reform since the Maoist era, but much less political reform."

China represents a problem for a democratic businesses—its political and moral cultures are repugnant, but its market is far too rich to ignore. "As businesses contemplate entering the China market and begin their processes of due diligence, most of them have actually already made up their minds: they cannot but be in China," Schell notes.

"Even for companies with the most noble of intentions, the unwritten laws of the free market do not provide a mechanism to reconcile the true cost of social responsibility with the fundamental need to be profitable," writes Karl Schoenberger in his book *Levi's Children: Coming to Terms with Human Rights in the Global Marketplace*. "An organization's instinct to succeed prevails over any lofty principles it might espouse."[7]

Google has not yet made this decision, at least not publicly. For years, Google has provided millions of Chinese citizens its service in the Chinese language, but as of mid-2005, it has yet to launch a subsidiary in China. That means that so far, the company has not had to play by Chinese rules when it comes to censorship of its main index. It also means that for the most part, Google has been left out of China's recent economic boom.

Regardless of its careful stance, Google already has a checkered history with the Chinese authorities. In the fall of 2002, the Chinese government began filtering out Google.com (and several other search engines) because those engines offered too many alternative routes to information that the government wished to keep hidden from its citizens. According to Chinese scholars in the United States, the loss of Google's service caused such a backlash among Chinese citizens that the government restored service within two weeks. Though it won't detail how it worked with the Chinese government to restore service, Google claims it was not forced to modify its service during the fracas—a claim that to this day, if true, makes it unique among major search engines. (After the shutdown, when Google users in China search for something that might return banned results, they see the links, but when they click on one, they are redirected to a government-approved site.)

But this was not to be the last time the company would wrangle with the China question. In early 2004, China came up again, this time in a more troubling fashion, at least from the point of view of those who wish to hold Google to a higher standard of good and evil. In February 2004, Google launched a Chinese language version of Google News. China immediately banned it—the site crawled a small number of news sources that the government found objectionable. Google immediately began negotiations with government officials, and the service was soon restored. But this time, Google purged the offending sites from its news index. Why did Google blink?

The company's official explanation was that to include the banned sites in Google's Chinese news index would create a poor user experience—when a Chinese user clicked on links from censored sites, he would find only error messages, and that would be frustrating. "Google has decided that in order to create the best possible search experience for our mainland China users we will not include sites whose content is not accessible," the company said in a statement, "as their inclusion does not provide a good experience for our News users who are looking for information."

But that explanation rang hollow to many—and worse, it sidestepped the real issue: by working with China to omit certain sites, Google had seemingly become an accessory to evil. After all, isn't it better to know that something exists, even if it is blocked, than to not know about it at all? Clearly Google was taking out all evidence of the banned sites because that's what the Chinese government wanted it to do.

The company initially refused to discuss whether this, in fact, was true. But the controversy began to balloon in the press and among influential blogs, and it became clear that Google was in danger of taking a major hit to its reputation. So the company released a clarifying statement, this time on its corporate blog.

For last week's launch of the Chinese-language edition of Google News, we had to decide whether sources that cannot be viewed in China should be included for Google News users inside the PRC. Naturally, we want to present as broad a range of news sources as possible. For every edition of Google News, in every language, we attempt to select news sources without regard to political viewpoint or ideology. For Internet users in China, we had to consider the fact that some sources are entirely blocked. Leaving aside the politics, that presents us with a serious user experience problem. Google News does not show news stories, but rather links to news stories. So links to stories published by blocked news sources would not work for users inside the PRC—if they clicked on a headline from a blocked source, they would get an error page. It is possible that there would be some small user value to just seeing the headlines. However, simply showing these headlines would likely result in Google News being blocked altogether in China. . . .

. . . On balance we believe that having a service with links that work and omits a fractional number is better than having a service that is not available at all. It was a difficult trade-off for us to make, but the one we felt ultimately serves the best interests of our users located in China.

Once again, this statement felt tortured—no one who understood how China works believed Google was censoring its news product for user interface issues, or even because of a desire to balance the availability of the service with what it termed "some small user value" of seeing the blocked headlines. Instead, it was clear Google had made an important policy decision to play by Chinese rules. Why?

The line "simply showing these headlines would likely result in Google News being blocked altogether in China" provides the answer. China is a huge market, and as a soon-to-be public company, Google could not afford to sit on the sidelines as competitors charged into the region. Yahoo, Microsoft, and others had already made their peace with the China question. But then again, none of them have adopted the motto Don't Be Evil.

And it turns out that something else was in play as well. In June 2004, news broke that Google had quietly invested an undisclosed sum in Baidu, the number-two Chinese search engine (the number-one engine, 3721.com, had recently been purchased by Yahoo).

Given the time it typically takes to consummate such an investment, and the fact that such transactions must be tacitly approved by the Chinese government, it is not hard to imagine the more substantial reasons for Google's decision regarding its Chinese news service—it didn't want to queer the Baidu deal, or any future moves it might want to make in China, including opening a subsidiary.

A minority investment in Baidu is one thing, but to truly prosper in the massive market, Google must run its own subsidiary, much as Yahoo does. Looking at it from a purely economic standpoint, the decision is obvious: if you are a major public company and there is a huge market opportunity, you must invest in it. On the other hand, if there were one company at this exact moment in history that might make a statement to the world that it will stand against the totalitarian regime of China, who better than Google? After all, this is the company that refused to sell banner ads during the height of the dot-com craze, the company that has maintained its moral ground and adopted a motto that—should it forgo China—would give it considerable air cover.

The China question weighs heavily on the conscience of Google's founders. Beginning in mid-2004 and continuing into 2005, Google began summoning the world's foremost experts on China to its Mountain View campus. According to several who were privy to these meetings, Google had one question on its mind: how can we go into China and yet not be evil?

"They can't afford to not be in China," says an eminent Chinese expert who spoke with Google's founders about the company's dilemma. "They are facing a hard choice. They really don't want to be seen as doing something that is evil, but no one goes into China on their own terms."

According to the scholar, Sergey Brin told him that were it up to him, they'd forgo China, but that he can't hobble Google's ability to grow. In China, Google may have finally found a situation in which its Don't Be Evil motto cannot stand.

"We look at China with a different point of view," Brin tells me during the internationalist World Economic Forum in Davos, Switzerland, in early 2005. "A lot of companies would say 'It's a big market. How do we get a chunk of it?' We want to focus on how do we do the most good."

On the one hand, Brin says, not having Google at all would be a disservice to all Chinese users. On the other hand, a censored service does run counter to his sensibilities. "You have to weigh the odds. Corporations need to be responsible. If we wrote [the Chinese laws] then I would say we were responsible for them."

But what of people who feel that Google is failing their expectations by not standing up to China? "I am sure at various times, various people will say that we failed their expectations," Brin says. "I think it's a good motivation to have, and I am sure we will not be perfect to everyone at all times."

"If you are manufacturing electrical lights, running shoes, cars, tools, or toys, all that really matters is cost. The bottom line rules," points out Schell. "The 'brand' suffers almost nothing by being 'Made in China.' But for a company whose product is something more intangible like knowledge, or even news that depends on freedom of access, the wager is, of course, somewhat different. This is all the more the case when the company is one like Google, which was not only born out of the IT revolution, but whose corporate persona is tinged with all the ideology of the early part of that revolution when values like freedom, spontaneity, independence, and resistance to control were some of the hallmarks of the new movement."

To make matters worse, should Google decide to capitulate in China, such a move could lead to charges that the company has done the same in any number of other places. "What may be most important is not the single concessionary act to China, but the precedent that this act would set for Google, namely, that the level of censorship before entry in specific markets will be negotiated on a case-by-case basis," Schell concludes. "If China manages to wring out such concessions, why should not another country or even some large multinational corporation which does not like unflattering information about it flying around the Google search universe, complain—and expect concessions?"[8]

It is odd to think that seven years after they started a company to "organize the world's information and make it universally accessible and useful," Brin and Page find themselves pondering a role as the morality police for the global economy. And it's doubly odd to think that the decision they take—whether to go in or not—will have a significant impact on literally billions of people's lives, not to mention untold billions of dollars in economic value. Certainly any number of large and important companies face conundrums like the China question, but Google sees itself as a different kind of company, one that makes its own way and refuses convention almost on principle. Nowhere would its unconventional approach surface more dramatically than when the company finally made the decision to go public in the spring of 2004.

Chapter 9
Google Goes Public

Success and failure are equally disastrous.

—Tennessee Williams

Sergey Brin is jet-lagged; he has the vaguely disoriented look of a young man still finding his bearings after a very long, strange trip. I watch him enter a crowded restaurant and look around for familiar faces—save for me, the persistent author, there are few. He is in Davos, Switzerland, attending the World Economic Forum (WEF), the annual conference of political and business leaders. The room is full of captains of industry and members of the media from around the world, and all of them stop to regard Brin, who is, quite literally, the man of the moment (he is slated to give a short dinner presentation that night).

Brin forges ahead around the tables, acknowledging a greeting here and there, his hands pressed together at his chest like a yogi's, his eyes more alert as he warms to the task at hand. He sits down at a table near the back, shakes hands all around, then informs his dinner companions that he really did just step off his plane. He was here to stand in for Larry Page, who was supposed to be at the dinner, but Page was feeling under the weather after the ten-hour flight.

It is January 2005, and Brin is at Davos for the fourth time, but this is his first as a billionaire helmsman of a newly public company. At last year's soiree, Bill Gates, CEO of Microsoft, acknowledged quite publicly that "Google kicked our butt" in search, but promised that Microsoft would respond with an even better offering. One year later, Microsoft had indeed introduced an early version of its new search software.

Back at the dinner, Brin is accepting congratulations and plaudits for Google's unusual initial public offering. The stock's stellar performance since the IPO (it had more than doubled in less than four months), had nearly everyone asking Brin what might be next for Google. Brin accepts the plaudits, but is clearly uncomfortable lingering on the story of the IPO itself. "We have more time to focus on the company now," he later tells one well-wisher. Clearly, Brin is glad the IPO is behind him.

The journey from dorm rooms and Burger King takeout to private jets and a starring role at the World Economic Forum has been dizzyingly brief; certainly Brin can be forgiven a resultant touch of jet lag. And as years go, 2004 ranked as a critical turning point for Google, the company, as well as Brin and Page, the men. For 2004 was the year Google began to grow up, not necessarily because it wanted to, but in the end, because it had to.

Rumors of an IPO

On October 25, 2003, the top story on news.google.com read: "Google Sparks Hope of New DotCom Boom." Given that the Google News computers choose stories based on popularity and prominence of source, it's fair to say that the speculation about when and if Google would file papers to become a public company had reached fever pitch. Later that same month, the *New York Times* reported that Microsoft was eyeing an acquisition of Google, a story that Bill Gates later disputed. In any case, it was clear that by the end of 2003, Google was crowned Silicon Valley's latest golden child.

Expectations were high—reports claimed Google's IPO would value the company at $16 billion, roughly the same size as Amazon.com.

As 2004 dawned, Google had become the talk not only of Silicon Valley, but of Wall Street as well. Whispered financials for the secretive company pegged 2003 revenue at nearly $1 billion, with profits estimated at more than $300 million.

By this time, both Yahoo and Microsoft had realized the threat Google posed to their businesses. Each of those companies had valuable public shares and massive piles of cash, and they scrambled to redeploy them against Google. Simply put, if Google was going to compete, it could not afford to stay private. Valley watchers, press pundits, and Wall Street writhed in ecstatic speculation: Would Google's IPO augur the second coming of the Internet bubble? Could it usher in a new, more profitable era of tech growth? Who would get rich? Who would fall behind? Who would follow in Google's footsteps? Might the company stumble?

In its early years, the company had downplayed talk of an IPO—after all, the markets were in the tank, and no one seemed to have an appetite for any kind of Internet stock, no matter how robust the company might be. But 2004 marked a transition of sorts—it seemed to be springtime again in the Valley—and the spotlight was squarely on Google. With its venture backers, its thousands of option-holding employees, and its massive profits, clearly the company was heading toward one of the largest public offerings in the history of technology. Right?

In fact, the answer was a qualified no. In an interview with the *San Francisco Chronicle* in the fall of 2001, Eric Schmidt laid down what would become the triumvirate's standard answer to the IPO question. "The IPO question we've debated internally, but frankly, we're profitable," Schmidt said. "We're generating cash. We don't ever need to go public."

This line was repeated, over and over, for the next three years, to the point where Google's evasive responses were becoming something of an industry joke. At a conference in early 2004, Brin even went so far as to joke that an IPO was not in the offing because "filling in all those accounting forms is too difficult."

Turns out Google's leaders were wrong about not needing to go public. Because the company had given stock options to more than one thousand of its employees, an obscure SEC regulation would force Google to begin reporting as if it were a public company, as early as April 2004. The stage, therefore, was already set.

Despite the realities of SEC regulations, that Google would become a public company was never really in doubt. Once a company takes money from venture capitalists, the event is nearly a fait accompli—only an acquisition or bankruptcy can easily divert the path. "The day I was hired I understood the company would go public because it had venture investors. The only question was timing," Eric Schmidt told me after the IPO, giving the lie to three years of transparently disingenuous corporate line-toeing.

But despite their company's obvious course, Brin and Page struggled with the idea of becoming public. Google had prospered in private, and its founders worried that the company would be forced into a mind-set of short-term thinking, a trait common to many listed companies.

Throughout 2003, Google toyed with scenarios that would allow the company to stay private. It hired consultants to model complex financial mechanisms—such as repurchasing options and the deployment of a shadow equity plan that might protect the company from avoiding its seemingly predetermined fate. But the math never satisfied Page, Brin, or their board—any way you cut it, the maximum payout for Google's investors was the public markets, plain and simple.

Meanwhile, Google had plenty of things to keep it busy. It was readying its Gmail application, as well as orkut.com, a social networking application meant to compete with the high-flying Friendster. (Google attempted to buy Friendster in early 2004 for $30 million to $40 million, but was rebuffed. Friendster later accepted venture funding but has since drifted off most Valley observers' hot lists.) An IPO would be a major distraction from ongoing development, and it wasn't as if Google's competitors were standing still.

As if to highlight that the company was still the search leader, in February 2004 Google announced it had increased its index size to 6 billion items, and it made a point of offering Brin to major newspapers to ensure the increase was covered.

But by early 2004, the buzz inside the Googleplex was palpable—employees were quietly told that the company was going to file for a public offering.

Google had been talking to several major investment banks, as well as to WR Hambrecht, a smaller boutique that specialized in auction-based IPOs. In traditional IPOs, a company puts itself in the hands of an investment bank, which determines the company's value and stock price—a process that many entrepreneurs believe favors the banks. Often, investment banks will price an offering below what it might receive on the open market in order to engineer a "pop" in the stock price. The bank then distributes pre-IPO shares to its favored clients. When the shares pop on opening day, the bank's clients reap huge windfalls. The company, however, has left money on the table—it sold its shares at the opening price, not at the top of the pop.

WR Hambrecht specialized in a new, more democratic approach to IPOs that uses a public auction to set the price of the stock before it becomes public. This alleviates the opening day pop, and, theoretically, garners more money for the company on the day it goes public. Using an auction process felt consistent with Google's nonconformist style, but it was not certain that Google's venture backers would support such a move.

Thousands of Google employees, spouses, contractors, and competitors began what would become an eight-month parlor game of guessing what the company would be worth and, more important, what their own holdings might come to. Word leaked out, and the parlor game turned into the Super Bowl of speculation—could this be the largest in the history of the Silicon Valley? Would Google go with an auction process? Would Wall Street really let Google be

Google? What would Google do with all that money once it was public?

The lessons of the past were not far from many Googlers' minds. On the morning of January 20, 2004, Google employee Eric Case, an engineer, posted a brief note to his personal weblog. Without comment, he quoted the musings of former Apple employee Bruce Tognazzini.

In the cold, early morning hours of a winter morning in 1980, Apple Computer went public. By the end of that frantic day, 64 people had become millionaires. I was one of them. Had I locked those stock certificates away in a safe deposit box that day, they would now be worth more than 18 million dollars. Instead, I "put them to work." Within 24 months, I had less than $300,000 left. . . . My ostensible purpose in writing this rather embarrassing treatise is, with luck, to prevent others from following in my footsteps.

"There are lessons to be learned there," Case later told me. "I think it was still my first week or two as a contractor where I woke up and realized if I were all of a sudden insanely wealthy, I'd still come into work every day."

But as Case and others would soon learn, it would prove difficult for employees of Google to hold back from selling their shares in the aftermath of the IPO. After all, when you are holding options on shares worth $200 each, and the opening price is $85, how crazy do you have to be not to sell?[1]

An IPO for the Ages

On April 29, 2004, Google filed what certainly had to be the most unusual S1—the formal public offering document—in recent memory. At filing, Google declared it would sell $2,718,281,828 worth of its shares—a seemingly random number, which was, in fact, the mathematical equivalent of e, a concept not unlike pi that has unique characteristics and is well known to serious math geeks. By manipulating the actual offering to provide this knowing wink to nerd humor, Google was in effect declaring: *the geeks are in control.* It would be the first of many such statements, starting with the rather startling news that Google would forgo traditional approaches to marketing IPOs and instead rely on an untested and modified version of a process known as a Dutch auction to distribute its shares. (WR Hambrecht did not lead the auction, but it did comanage the deal with a number of other, more traditional banks.)

The S1 ran well over one hundred pages and kicked off with a letter to prospective shareholders, penned by Larry Page and titled "An Owner's Manual for Google's Shareholders." In it, Page outlined how he, Brin, and Schmidt intended to run their company. The letter also served as a manifesto declaring what Google was really all about, a statement by the founders of their company's role in the world. Given the quiet period imposed by the SEC on all companies during the process of a stock offering, the letter served as the founders' single chance to define themselves in the eyes of the world. It didn't disappoint.

Personal, discursive, and sometimes defensive in tone, the letter attempted to address an investor's most pressing questions. It claimed, several times over, that Google was different, special, and remarkable. It also acted as something of a caveat, a pardon for future sins, claiming that going forward, Google would not act the way public companies are supposed to act, because it was unique. "We're different and better than others," was the tone. "We know best." Page's first sentence sums it up: "Google is not a conventional company. We do not intend to become one."

The letter made more than a few observers cringe—on Wall Street and beyond. From Wall Street's point of view, the letter was nothing short of a defiant middle finger. Inspired by renowned financier and folksy Wall Street hero Warren Buffett, whom Page cited in the letter, Google announced that it would retain an unusual amount of control over its newly public status. "The standard structure of public ownership may jeopardize the independence and focused objectivity that have been most important in Google's past success and that we consider most fundamental for its future," Page wrote. "Therefore, we have designed a corporate structure that will protect Google's ability to innovate and retain its most distinctive characteristics."

In the letter and elsewhere in the S1, Google outlined a "dual class" shareholding structure, one in which the founders and senior executives hold far more control than the common shareholders. In essence, while Page and Brin would jointly hold just 30 percent of the actual shares in the company they founded, they nevertheless have control over every major decision the company faces, because each of their shares holds ten times the voting power of those they intended to sell to the public.

Such dual class structures are rare in public companies, but common in media companies that are family owned, including the Washington Post Company and Dow Jones, owner of the *Wall Street Journal*. "The main effect of this structure is likely to leave our team, especially Sergey and me, with significant control over the company's decisions and fate," Page wrote. "While this structure is unusual for technology companies, it is common in the media business and has had a profound importance there. . . . [D]ual class ownership has allowed these companies to concentrate on their core, long-term interest in serious news coverage, despite fluctuations in quarterly results."[2]

Page and Brin had even more unusual plans. Besides choosing an auction process and dual voting class, Page announced that Google would not be providing Wall Street with traditional earnings guidance, and that furthermore, Google would not attempt to "smooth" its earnings to create the impression that the company was on a stable and steady path of growth. Reinforcing the company's unconventional approach, Page outlined how he, Brin, and Schmidt run Google as a triumvirate, sending a very clear message that Schmidt, while a key player, was by no means the final word on any corporate decision.[3]

To summarize, Google pretty much flouted traditional Wall Street approaches not only to selling shares, but also to corporate governance, investor communications, and management structure. Not surprisingly, Google's filing began a period of decidedly mixed press accounts—partially because the company could not make its case, owing to quiet period restrictions, but also because any number of Wall Street types were more than happy to take the company down a peg or two in retribution for its perceived arrogance.

"I didn't realize it would be such a big deal," Brin later told me. "Seriously."

The founders may not have realized the shit storm their approach would stir up, but Google's venture capitalists certainly did, and according to several sources close to them, they were not happy with the founders' insistence on flouting Wall Street tradition. "I think our attitude is 'Let's not be too cute,' " one venture capitalist told the *New York Times* in the week leading up to the filing.

Google also came under some withering criticism from technology-industry veterans, graybeards who had seen countless companies go public, and felt the Google guys were perhaps getting high on their own supply. "Google wants to have its cake and eat it too," Mitch Kapor, founder of Lotus and noted Valley investor, wrote on his weblog. "Google says: Give us your money and we'll sell you a lottery ticket. We know what we're doing, so it would be counter-productive for you to have any control over what we do. Sit in the backseat and enjoy the ride and don't think too much about the odds."

Others were even more reproachful. An entrepreneur in the search industry e-mailed me:

> What a hugely immature and ego driven thing to do. . . . To pretend that this short-term success somehow is due to (or indication of) some hugely revolutionary business thinking by Page and Brin is pretty bizarre. I can only think the lawyers let them leave it in because they look forward to seeing it get thrown back in their face when they come down the backside of the current Google

> worship curve. And there certainly is a long way down from where they are right now.

But in the end, money talks. Besides Page's controversial letter, the S1 also included the first ever glance at Google's financials. And not to put too fine a point on it, they were extraordinary. Google's first public income statement showed that profits were on track to break a quarter of a billion dollars in 2004, and that the company had made more than $100 million in 2003. Analysts quickly noted that those profits were, in fact, depressed by various accounting requirements, and that the company had generated more than half a billion dollars in cash in 2003 and was running at operating margins of more than 60 percent—"stunning" according to Mitchell Kertzman, a venture capitalist quoted in the *Wall Street Journal.* In terms of financial metrics, Google was proving that it was indeed a very different kind of company.

A Rocky Offering, a Rocket Ship Afterward

The offering took longer to complete than anyone had imagined. Preparations were confounded by several factors: the company's uneven management of its own overwhelming growth, the relentless and distracting scrutiny it was suddenly facing, and, internally, the founders' continued reluctance about the public path—according to Schmidt, Page and Brin were not sure about going public until the very day the stock opened on the NASDAQ exchange in August 2004.

The combination of these factors worsened Google's reputation as a partner in the eyes of many. Whereas before it was simply difficult to get information and responses from Google, now it was damn near impossible. Google's penchant for secrecy increased to nearly paranoid levels. Employees were warned that any slip might kill the deal—and no one at Google wanted that to happen. While the company's culture discouraged open discussion of wealth, certainly everyone there—roughly two thousand employees by that time—was quietly counting his chickens as the hatching point neared. More than half of the employees were set to become millionaires. No wonder the company circled its wagons.[4]

"I recently went there to talk to some folks about an idea I had," a seasoned Valley entrepreneur told me after a visit in the summer of 2004, a month or so before the IPO. "I came out feeling like I had visited a fascist state. It's as if everyone there feels lucky to be there, and they have dummied up—no one wants to say the wrong thing."

Beyond that, Google had a lot of cleaning up to do. The company was not prepared for the rigors of being a public business, in particular the strictures of the Sarbanes-Oxley Act, passed in the wake of the corporate scandals that had rocked the United States. Among other things, the act tightened rules that concerned accounting for revenue. While not as onerous for companies that make their money thousands of dollars at a time—automobile manufacturers, for example—it was hell for a company like Google, which made its money literally pennies at a time, from millions upon millions of microtransactions. According to engineers involved in the work, Google had to significantly restructure its advertising reporting system from the ground up.

Such a project meant that the Neil Moncriefs of the world—small advertisers with significant grievances—found Google less responsive than ever. Here was a company that was aiming to reap nearly $3 billion from the public markets, but it still didn't have time to answer the phone.

As the summer wore on, speculation ran rampant among many in Silicon Valley that the markets would hand Google a long overdue comeuppance. And the naysayers had a point: for the past three and a half years, the technology IPO window had been pretty much nailed shut. Amendments to Google's S1—viewed as milestones in any IPO's progress—were slow to come, and rumors began to surface that the company was having trouble with the technology behind its unique auction process. Furthermore, August loomed, a month when much of Wall Street is on vacation.

The press jumped on the IPO's lack of progress, and a steady drumbeat of articles began to question whether Google's unusual filing would live up to the hype that preceded it, and whether the company could maintain its folksy approach to business given the realities of Wall Street.

"The real question is whether Google, like Buffett, will be able to ignore Wall Street's demands and go its own way," wrote Allan Sloan, the Wall Street editor of *Newsweek*. "I doubt it. . . . Google will have to pay attention to its stock price—and thus, to Wall Street. I love the way that Google dissed the Street in its filing—distrusting the Street is the right move. Going public, I fear, will prove to be the wrong one."

Adding insult upon insult, Google's management was hit with a Google bomb—an intentional attempt to manipulate the results of a search so as to discredit someone. In June 2004, typing "out-of-touch executives" into Google returned the biographies of Google's top management as the first result.

Google was stung by the bad PR, but given the quiet period it had little recourse. It did launch a corporate blog in May, but the site proved sterile. Put simply, Google had to grin and bear it. At an industry conference in early summer, Eric Schmidt was seen walking around wearing a T-shirt that read QUIET PERIOD on the front, and CAN'T ANSWER QUESTIONS on the back.

By late July, Google had chosen Morgan Stanley and Credit Suisse First Boston as its lead banks and indicated that it had picked the NASDAQ as its exchange. The company also announced its price range for its stock: $108 to $135—extraordinary, as most companies try to price their stock in the teens so as to attract retail investors. Google could have split its stock in order to bring the price down, but refused to do so. The news brought a fresh wave of negative coverage—Google was accused of pricing out the little guys, the very investors its auction was intended to empower.

Unable to respond, Google pressed on. Rumors had it that the company would go out by the end of July. But Google managed to continue business as usual, making the Picasa purchase, for example.

However, the company managed to take yet another hit to its reputation when Brian Reid, a former senior manager, sued for age discrimination. And the SEC then announced that it was recommending civil action against David Drummond, Google's general counsel, for accounting irregularities involving a company he worked for prior to Google. The news, it seemed, was only getting worse. "Why Not to Bid on Google IPO" ran one headline in the *San Jose Mercury News*. "Google's IPO, Asking Too Much?" asked *BusinessWeek*.

In late July, Google opened a Web site where the public could register to purchase shares, and the triumvirate began its road show—a presentation given to institutional investors in advance of a public offering. Unfortunately, reviews for those appearances were mixed. "They were really unprepared," says one investor who was at a presentation in New York. "They didn't seem to be ready for the questions they were getting." Other investors told me the Google guys did fine, but were clearly sitting on their hands, trying as hard as they could not to hype the company lest the SEC slap them back.

The SEC did slap them back, but not for overhyping the company. Instead, the SEC reprimanded them for offering millions of shares to their employees that had not been registered with the SEC, an offense which made Google's management look like bumblers and forced the company to conduct what is called a recision offer—a legal process in which it had to offer to repurchase the shares at their value when they were first offered. (Given that the stock was worth far more in the present than at any time in the past, no one took Google up on the offer.) In the end, the recision offer was not a major setback, but it certainly didn't help the company's image—the Google IPO was not going well, and here was yet another example.

The news kept getting worse. In early August, the *Wall Street Journal* reported that glitches in Google's auction technology had indeed delayed the offering. Reports from nearly every major newspaper claimed that Google was not hearing good feedback from institutional investors, any number of whom were more than happy to be quoted declaring they intended to sit out Google's auction altogether. (Of course, it benefited those same traders to claim this—

if it drove the stock down, they could submit lower bids and win the stock at a discount.) Rumors began circulating that Google would have to drop its offering price as a result.

And then the *Playboy* interview hit. Back in April, one week prior to filing Google's S1, Brin and Page had given an interview to *Playboy*. According to Google, the magazine had promised to hold the interview until "the fall." From Google's point of view, that felt like a date well past the IPO. But to *Playboy*, eager to publish its scoop, fall meant September, and given that most magazines hit the newsstands about a month before their issue date, September really meant August.

Most anyone in public relations will tell you that giving an interview to a major publication one week before filing an IPO is a mistake. Google, it seemed, was determined to sabotage its own IPO. The interview itself was relatively harmless, but the founders did make a number of claims which contradicted facts in the S1, including the number of employees at the company (they said "about 1,000," but the actual number was more than 2,200), and the number of visitors google.com receives (the article reported it as 65 million a day; Google later clarified it to 65 million a month). On Thursday, August 12, the SEC announced it would investigate the interview in order to determine whether it violated the quiet period. To appease the SEC, Google entered the entire text of the article, along with some clarifications, into its S1.

To make matters worse, the markets themselves were falling apart. The NASDAQ, which had peaked for the year in January, slid below 1900, and the mood on Wall Street was deteriorating. Several Internet-related IPOs filed in anticipation of a "Google lift" were instead pulled, leading many to conclude that Google had no choice but to pull its offering as well.

Tempting Fate

Anyone waking up in a black mood on the morning of Friday, August 13, 2004, could certainly be forgiven. There was plenty of bad news to support the superstition that Friday the thirteenth is an unlucky day. World oil prices were spiking, prompting many analysts to warn that a global economic recession was in the works. The stock markets were in the dumper. Opening ceremonies for the Athens Olympics were slated for the evening, but news of the event focused mostly on terrorism—the media seemed convinced that Al Qaeda was conspiring to attack the games, and the opening gala seemed the perfect venue.

So of course, on the unluckiest day of the year, after several major setbacks and in the worst market since the dot-com bubble's bursting, amid an oil spike and threats of global terrorism, Larry Page and Sergey Brin decided to press ahead with the process of auctioning their shares to the public. Friday the thirteenth was not the day Google first traded on the NASDAQ—that would come a week or so later. But it was the first day of an auction that would set the company's initial price, and therefore, the beginning of Google's life as a public company.

What were they thinking?

Surely Brin and Page weren't thinking about the Olympics as they labored through the process of bringing their company public, but the timing was noteworthy just the same. Scheduled for Athens, Greece, in a nod to the one-hundredth anniversary of the modern games and their origin in Homeric times, the 2004 Olympics were riddled with delays, cost overruns, and suffocating fears of terrorism. Launching the games on what is understood to be the unluckiest day of the Western world's calendar was courageous, to say the least.

But did Google need to do the same? Its offering also suffered numerous setbacks and delays, and was the most expensive technology offering in recent history, in terms of share price. Certainly no one would have blamed it for waiting until the following Monday—not the querulous bankers on Wall Street, bested only by baseball players for their superstitious leanings. Should there be an act of terrorism on Friday night, the markets would tank on Monday—all hell might break loose. Why not wait a day and see what happens? *How can those guys tempt fate so baldly?*

In fact, the Greeks believed that your fate is already sealed—you can no more tempt it than calculate the final digit of pi.[5] In engineering terms, fate is a mathematical proof. Your free will to chose this day or that for your IPO will, in the end, have nothing to do with your ultimate fate. This whole notion of tempting fate is a bagatelle created by men terrified of math: the result, in the end, is simply the result. Damn the torpedos, full speed ahead!

The opening ceremonies for the Athens games went off flawlessly, but the same could not be said of Google's auction process. After a few days of watching the bidding, Google executives and their bankers realized that the stock would never price in the (already quite broad) range they initially had chosen: $108–$135. The auction was delivering nearly perfect market information, and the market was giving Google's shares a serious haircut. On August 18, the company announced it was cutting the range of its offering price to $85–$108, and lowering the number of shares it would offer to the public by 6.1 million. The bad news, it seemed, would never stop.

Lowering the range happens all the time in iffy markets—often it's a sign that the offering is in trouble. In the weeks prior to Google's offering, several other companies had also lowered their range. But in Google's case, there was an additional factor: the nearly perfect window of market demand information. Armed with that, Google's managers could more accurately predict what would happen in the aftermarket once the offering went live, thereby allowing them to lay out scenarios for several potential chess moves, and make the best decision on price. Perhaps Google could have gone out within the original range, but if it had, the stock might have dropped significantly in the aftermarket. The only people who really cashed in would have been insiders and the company itself—ordinary investors would have been soaked.

On August 18, Google formally asked the SEC to approve its offering, even with the pending investigation into the *Playboy* article (the investigation was eventually dropped). Perhaps sensing that Google had been through enough, the SEC complied. On August 19, nearly four months after filing its initial prospectus, Larry Page rang the bell on the floor of the NASDAQ (Brin stayed back in Mountain View with the troops), and Google Inc. finally went public—at a price of $85 a share.

What happened next put to rest nearly every doubt about Google's offering. By the end of the day, the stock had rocketed to nearly $100. By the next day, GOOG was at $108.31—breaking into its original predicted range. And Google kept climbing, topping $200 by November.

The extraordinary performance of Google's stock was fueled by more than just hype. The company's first quarterly report as a public company showed sales doubling from the prior year. Wall Street analysts subsequently praised the company for its execution and market strength, and the stock held its lofty position near $200 from that point on. It didn't hurt that the overall market in online advertising was growing faster than any other sector in the media business, of course, but Google also delivered on its promise to keep innovating, announcing a steady stream of new products in the months after the IPO. After its second quarterly report exceeded the first, influential analyst Safa Rashtchy of Piper Jaffray raised his price target for the stock to $250. The stock climbed to nearly $300 by early summer. The fates, it seemed, had been smiling on Google after all.

Now What?

"I am not superstitious," Eric Schmidt tells me a few months after his company's IPO. I have been asking him about launching his offering on Friday the thirteenth. "My job was to land the airplane. We were on a turbulent flight. As long as we got all the passengers off the airplane and we're safe, I'm happy."

But what of all the missteps, the terrible press, the *Playboy* interview? "The *Playboy* interview was a low point," Schmidt acknowledged. I then asked him about his relationship with Wall Street. After all, he was the CEO; he signed the financial statements. Surely he had some patching up of relationships to do with the Street? As far as Schmidt was concerned, the answer could be found in his company's numbers (Google had since announced two quarters of extraordinary earnings). He then went into something of a rant, one too good not to replay:

"The ninety-nine percent important thing to do in a company is to win, OK," he started, a bit agitated. "The one percent is to be very, very nice while you're winning." Referring to his prior roles at Sun and Novell, Schmidt continued: "Now I have been ninety-nine percent nice as I lost and lost and lost and it gets to you a little bit. The fact of the matter is, Wall Street is a performance organization and they care about winning. I am held to that standard and I think that's appropriate. . . . I am sure that there are people who were negatively affected by the decisions that we made and there was no disrespect meant. We behaved honorably; we chose our path. If you didn't want to participate, it's OK. Please take your money elsewhere. There are many, many ways to run the world, run your company, run your IPO. One of the things that bothers me about modern culture is that there is a presumption of only one correct way of doing things. It's perfectly fine for Google to be run the way we run it as long as we tell you the truth. If you don't like it, *don't participate*. You're here as a volunteer; we didn't force you to come. Right?"

Was he upset that Google took so many hits during the run-up to its offering? "I'm upset that we became four months' worth of IPO football," Schmidt admitted. "But now it's important that we go back to doing what we are all about."

So what *is* Google all about? Now that the company had had its IPO and could get back to work, the question still hung in the air: what would the company do next? With more than $3 billion in the bank and a market cap pushing $50 billion, clearly the company needed a plan. While traditional companies—some might call them mature—have well-understood corporate development plans, the post-IPO Google was still flying by the seat of its pants. How long could the company keep it up?

Chapter 10
Google Today, Google Tomorrow

He who has great power should use it lightly.

—Lucius Annaeus Seneca

For the first few months after the IPO, Larry Page seemed to withdraw from his role at the company. He was already the more reticent of the triumvirate, and his colleagues reported that he now pulled back even more, refusing public appearances and press interviews. Certainly Page was used to being a rock star—Google had already made the cover of nearly every major magazine—but the cautious word from within the Googleplex was that fame, wealth, and their attendant responsibilities had gotten to the young founder. "He is really stressed out," said a senior executive with the company when I asked whether Page was available to talk in late summer 2004. "It's not a good time to be asking anything of him."[1]

By the time I met with Page in November 2004, he seemed quite composed—whatever demons had visited him after the IPO must have been brought to heel. But given how dependent Google was on his and Brin's leadership, I had to ask: Had he gotten used to his own level of fame and wealth? Had he adjusted to it as a person?

"I hope not," Page replied. "In a company like this, everything doubles in a year. Everything you did stops working. My job has always changed radically on almost a weekly basis."

But was he used to being a public figure? I pressed. "I'm not really used to it," Page admitted. "I just want to invent things and get them out into the world. I really feel lucky that I have the ability to affect things now. It's a tremendous responsibility to use that for good. . . . I feel more pressure to do things that matter. I'm responsible to a lot of people now."

As Google looks toward its own future, that responsibility—to shareholders, to employees, and to users—will only increase. Google faces perhaps its most tremendous test in the next few years—can it continue to innovate in the face of treacherous competition? Can it keep its most productive employees despite their own personal wealth? Can it learn how to partner with outside companies who find Google's loose approach to business confusing and dangerous? And finally, can the triumvirate of Schmidt, Page, and Brin hold it together—is it the right team to take the company from 3,000 people to 30,000?

Taking the next step as a public company required a bit of navel-gazing. Beginning in the middle of summer and continuing through the late fall of 2004, Google underwent a process of strategic review, starting with once again elucidating its core principles and values, then working out toward tactical questions: How should the company organize itself? What were the company's most obvious opportunities? What could the company do better?

"Virtually every issue that Google has is predictable," Schmidt told me, referring to his company's looming challenges. "Talk to anyone who has been through a high-growth phase and ask them what mistakes they made. We're making all the same mistakes. The question is, are we making them in an aggregate amount less or more? We have all the problems of growing from a small core group: strategy, buy in, motivation. How do we manage the issue of wealth creation, income levels; how do we compensate people with a high stock price versus a low stock price? It's a very long list. You have a better chance of succeeding if you stay focused on those problems. We have Microsoft coming into the market to compete with us, Yahoo executing very well."

Google embarked on a post-IPO strategy review for one reason: it was long overdue. Even Page and Brin, never fans of the traditional corporate process ("I'm not a big believer in strategy," Page once told me), agreed that Google had outgrown itself.

Leading the charge was Shona Brown, a longtime McKinsey consultant whom Google had hired as vice president of business operations. "Most of what I do is making sure we don't internally implode," Brown told me. "We don't want to fail because we don't execute."

Google's introspective phase began, as nearly every critical project at Google does, with Larry Page and Sergey Brin. The founders holed up for an all-night writing session and emerged with what became known as the Tablets—a definitive declaration of what makes Google Google. While Google won't divulge the contents of these sacred texts, Schmidt did qualify them for me. "They are very high-level stuff. Principles and values," Schimdt said, then laughed. "I said to Larry and Sergey—what am I supposed to do with this? I have the Tablets, and I have a lot of engineers."

Together with Brown, Schmidt took the Tablets and used them as a guide for a months-long management process that evaluated all of Google's practices. The result was a new organization—one that Brown says will allow the company to grow from its post-IPO size of nearly 3,000 employees to something "ten times that size."

Whereas before Google was run by a core group of senior managers responsible for everything, Google post-IPO is organized into a set of core groups by function—core search, advertising products, and what the company calls "20 percent" and "10 percent." These are designations for products that sprang from acquisitions or from the company's fabled product development process, whereby engineers are encouraged to pursue other interests beyond their core workload. (One engineer at the company joked to me, "I'm not quite sure when we're supposed to find that twenty percent—is it between brushing your teeth and going to bed?"). Ten percent time is for really wild ideas, things that, at first glance anyway, are difficult to justify against Google's current business lines.

Examples of "20 percent" items include Gmail, Google News, and Orkut. Ten percent items include Google's Keyhole product, a satellite mapping service that the company has integrated with its Google Maps product, and Picasa, a photo organizing tool.

In essence, the company has focused divisions executing on its two core businesses (search and advertising) and more loosely run groups pursuing projects that may or may not turn into core businesses along the way. This ain't exactly GE—Google's executives are still wary of becoming too rigidly organized—but they are trending that way.

As an example, Google's famous Top 100 list was dropped in late 2004. Having a centralized list of prioritized projects worked when the company was run from the center. But given Google's growth, "we've had to segment it," Schmidt told me. Now, each core group has its own list of projects to pursue.

As for Google's org chart, Schmidt takes nearly all the key reports, leaving Page and Brin free to pursue their own interests and agendas. But a new structure doen't mean that the founders aren't still firmly holding the reins. One day last fall, Schmidt found Brin sitting in his office in a Japanese massage chair, staring at his computer. Thanks to the particular properties of the chair, Brin was literally shaking in his seat (Brin has had to manage a low-grade back injury for years). "I asked him 'Sergey, what are you doing?' " Schmidt told me. "And he said 'I'm going through project by project.' "

According to Schmidt, there were at least five hundred projects across all the various segments of Google's burgeoning business, and Brin was reviewing them all. Even with the new structure, it seems the founders are still very much in control. (As an example of the founders' grip on Google's strategy, consider the music business. AOL, Microsoft, and Yahoo all have divisions focused on selling music over the Web. I asked a senior manager at Google why Google doesn't do the same. His answer? "Sergey doesn't listen to much music.")

Persistent Growing Pains

The question of how the three top leaders of Google interact will continue to fascinate Wall Street, but all three claim, as one might expect, that the triad is working well. As with all companies led by strong founders, an emperor's-new-clothes syndrome can set in, and many inside and out of Google claim the company suffers as much as it prospers from the cult of Page and Brin. Many complain that to get anything done—at least in the past few years—you had to get the approval of Brin and Page, and the two founders have only so much time in a day.

One such complainant is Brian Reid, a venerable Valley engineer who was recruited into a senior management position at Google in 2002 at the age of fifty-two. (As mentioned in Chapter 3, Reid worked at DEC during the AltaVista days.) Less than two years after joining Google, however, Reid was fired, and he subsequently has sued Google for age discrimination.[2] The text of his complaint paints an unflattering picture of Google's culture, but spend an hour talking to the man, and it quickly gets worse. Reid clearly has an ax to grind—he believes he was bilked out of millions of dollars' worth of options—but for a respected engineer with decades of experience to speak out so directly is certainly rare.

"Google is a monarchy with two kings, Larry and Sergey," Reid told me in an interview just before he filed his lawsuit (he has since refused to speak to the press). "Eric is a puppet. Larry and Sergey are arbitrary, whimsical people. . . . They run the company with an iron hand. . . . Nobody at Google from what I could tell had any authority to do anything of consequence expect Larry and Sergey."

Reid claims he was fired because he did not fit into Google's "youth-obsessed" culture. He also claims Google tried to keep his firing quiet, and pay him off with a severance package tied to a nondisclosure agreement that he refused to sign. Google will not comment on the Reid case, as it involves pending litigation.

Reid's comments echo many statements made to me—all on condition of anonymity—by Google's partners, competitors, and several past employees and contractors. However, all this ire must be taken in context. Because Google is an extremely important and powerful company, one driven by two charismatic and extremely bright young founders, it isn't hard to find people with nasty things to say. The same could be said of Larry Ellison at Oracle, Steve Jobs at Apple, or Bill Gates at Microsoft. I asked Eric Schimdt about the bile that seemed to be reserved for Google, and in particular Page and Brin. Was he surprised?

"You frame it as though it were a problem," was Schmidt's rather smug reply. "The beauty of Larry and Sergey is that they are well-known quantities, that if you don't want to work with them please don't. Slavery was made illegal years ago."

The trouble with Google, if it can be called trouble, is that the company rocketed from being unknown to having the status of Apple or Microsoft in five years—a rather unprecedented feat. In fact, the accounting firm Deloitte Touche named Google the fastest-growing company ever—noting that its five-year revenue growth exceeded 400,000 percent.

Such heady growth could kill almost any company. It requires an unusual combination of luck, brains, and hardheadedness to survive. And it's no wonder, in the end, that any number of people in Google's wake felt hurt, ill treated, or ignored.

Both Page and Brin acknowledge in interviews that they are exacting managers. And to be fair, far more employees at Google sing the founders' praises than grumble about their vagaries. As for his own role in the company, Schmidt says he makes the trains run on time, and leaves vision and product development to the founders. The team still employs Intuit founder Bill Campbell as a coach of sorts, but a company spokesperson downplayed reports that the company would be lost without his guidance.

In the end, however, Wall Street likes its companies to be run by a committee of one. In time, it's likely a single leader will emerge, as is the case at Google's two main competitors, Yahoo and Microsoft.

The Competition

Google's competitors are legion, but the most important of them all, at least in 2005–2006, is Yahoo. Microsoft, like an aircraft carrier lurching into a ten-mile tack, will certainly be a force to reckon with by 2007, but Yahoo is Google's main foe in the present day, and it is striking how similar, yet distinct, the two companies really are.

Two young Stanford PhD candidates as founders, one more gregarious, the other more withdrawn. Humble beginnings in a dorm room. A fascination with search and the vastness of the World Wide Web. A silly name that caught on and became cultural shorthand for the Internet itself. Extraordinary hypergrowth and success, marked by top-tier venture capital investment, a wildly successful IPO, and a multibillion-dollar market cap. Certainly Yahoo shares many key characteristics with Google.

But Yahoo is not Google, and the differences between them are illuminating. Consider the founders. While both sets of founders remain at their respective companies in important roles, Jerry Yang and David Filo, founders of Yahoo, are self-effacing, deflective of credit, and quick to delegate authority and responsibility to others. "Jerry is probably the most decent guy you'll meet in the Valley," says friend and investor David Siminoff, a well-known Valley financier (and admitted Yahoo partisan). "They let Terry [Semel, the CEO of Yahoo] run the company. But the Google guys, well, they rule with an iron mouse over there."

Siminoff's comments were reinforced by scores of senior Valley folk I interviewed during the course of reporting this book. When you walk the halls at Google, it's clear that Brin and Page are the bosses. Over at Yahoo, on the other hand, Filo and Yang are the founders, and therein lies the difference. It's hard to be a micromanager when your role is long-term vision and the CEO is a major force from Hollywood in his own right. Yang and Filo prefer to let Semel and his lieutenants speak to issues of corporate strategy on a day-to-day basis.

Inventory the campuses of Google and Yahoo, and again, one is struck by the similarities first. Both companies have built (or leased) headquarters that create a communal ambience. Both incorporate modern three- to six-story office buildings that surround grassy open spaces sporting basketball or volleyball courts. Both have spacious, if sometimes crowded, cafeterias that crank out an astonishingly healthy menu of culinary options for hundreds upon hundreds of young workers clad in jeans and T-shirts.

But at Yahoo, you have to pay for your lunch. At Google, lunch is free.

Why is there no free lunch at Yahoo? In 2001, Yahoo got smacked upside its head by the markets, and was nearly written off as a company. It had to lay off hundreds of workers, consolidate its cost base, and watch its stock drop from highs of more than $500 to lows of less than $10. Its employees—the ones that were left—walked around their campus in a state of shock, shoulders drooped, sapped of confidence. In short, Yahoo has seen the business end of failure, and has been chastened by the experience. But Google, well, Google has never known anything but success. The only thing Google has failed to do, so far, is fail.

Another distinction, according to entrepreneurs and advertisers who have worked with both companies, is that it is on average far easier to do business with Yahoo than it is with Google. Yahoo is four years older than Google as a company, and that fact alone may explain it—Yahoo's simply had longer to learn how to be a good partner. "At Google," one developer of a Web-based consumer service told me, "it's tiring to try to get anything done. It's chaos. No one knows where the meeting room is. Then the key people are forty-five minutes late. Then people are going in and out, with new people coming in every twenty minutes. You have to keep starting over, as the new people are not briefed on what the meeting is supposed to be about." Afterward, he continued, "nobody followed up, and when I called to see where things stand with our deal, I got sent to yet another group of people to start the same process all over again."

But when the same entrepreneur visited Yahoo, he found an entirely different experience. "Everyone was on time and they had read up on my company, and knew what they wanted from the meeting. It lasted exactly one hour, and the follow-up was clear and focused."

That's a function of experience, but it's also a function of culture. Success and hypergrowth breed a certain level of arrogance and insular thinking in any company. There's no dearth of stories about the messiness of Yahoo's business culture circa 1998–1999, but those stories pale compared with the level Google had achieved by the time of its IPO.

Google isn't dumb; it was aware of these problems even as it continued creating them. In 2003, it hired Megan Smith, cofounder of Planet Out and generally one of the more beloved people in the Valley, to help it run its business development department, and Shona Brown continues to plug away on issues of business process. I asked Brown where she thought Google had improved the most since she arrived in 2003. Without missing a beat, she identified partnership. "We are much more open and less insular," she said. "By that I mean we are working much better with our broadly defined sets of partners. We realize we are part of the ecosystem and have to work with others. That has been a huge and positive switch."

Paging Usher

Google and Yahoo differ in more than just business culture. They also differ in approach to the core application that drives profits, search. Consider a search, done in late 2004, for the one-word term "usher." Further, let's presume the person typing that search in reality does want to know about the popular singer by the same name.

On Google, "usher" brings you a pretty predictable set of results. Because Usher the singer is quite popular and therefore much in the news, Google incorporates some Google News stories into its results. On the right are plenty of AdWords related to Usher—there is no shortage of vendors who stand to make a buck or two off the man. The majority of the page, however, is given over to listing Google's top ten results for the keyword.

The first three results, starting with UsherWorld, are clearly relevant to the keyword entered, again assuming that we are looking for information about the singer. The rest of the first page of results mixes in Edgar Allan Poe's "The Fall of the House of Usher" as well as the usher syndrome, a rather obscure communication disorder. It seems some kind of diversification algorithm is at work behind Google's curtains—if the engine chose purely on popularity and links, the first few hundred, if not thousand, results would most likely be about the singer.

But in terms of exploiting our intention behind the search term "usher," that's as far as Google goes. Save Google News, the company offers very little overt editorial guidance. You're directed to Usher's Web site, and that's that.

In contrast, consider how Yahoo handles the same search. "Usher" on Yahoo Search also gives UsherWorld as the first organic result, but the similarities end there. The first thing you see below the search box is Yahoo's "also try" feature, asking if you, the searcher, might be looking for a more refined version of an Usher search. Perhaps you're looking for lyrics to a particular song ("usher lyrics" or "usher my boo lyrics"), or for pictures of Usher ("usher pictures"), or for more information on the star's relationship to Alicia Keys ("usher alicia keys"). This feature is driven by Yahoo's editorial decision to watch what its users are searching for and connect the patterns it sees. Behind the curtain, Yahoo makes lists of related searches, then surfaces the most relevant ones. According to conversations I've had with members of Yahoo's search team, the "also try" feature is a huge hit with Yahoo users.

Below "also try" are two blue-backgrounded sponsor results, right at the top (there are also plenty of paid links to the right, as there are with Google). This reflects Yahoo's more aggressive approach to commercialization throughout its site. In all my discussions with Yahoo executives, I've noticed a distinct pride when it comes to commerce: integrating commerce directly into the search process is seen more as a benefit than as a detriment. The premise is that search advertising is in fact relevant and even helpful to a searcher (a premise that, to be fair, is also echoed at Google, but in an almost apologetic fashion).

The practice of listing sponsored results right up at the top of the page occurs in more searches on Yahoo than it does on Google, but it does happen at Google: a search for digital cameras or American Blinds, for example, brings paid listings to the top of Google's results. (In court transcripts in the American Blinds case, Google's lawyers assert that the practice of putting paid search results at the top, which many claim is confusing to users, has ceased at Google, but it clearly persists, if in more limited fashion.)[3]

Continuing with Yahoo's results, we next see a vital distinction between the ways Yahoo and Google handle the intent of their users: Yahoo's search shortcut. The shortcut is Yahoo's attempt to bring all the most pertinent information about Usher into one place at one time, so as to quickly allow the searcher to declare and execute his intent. In four or so lines, the shortcut result offers the Usher artist page on Launch (Yahoo's music service), photos and videos of the artist (also on Launch), and the ability to buy the artist's CDs (on Yahoo Shopping). Yahoo News results are incorporated as well. The entire shortcut is flagged by a small red "Y!" so the searcher is tipped off that this particular result comes from Yahoo's own editorial judgment, rather than the Web.

Last come Yahoo's organic results. It's interesting to note that with Yahoo there is far less diversity in the first ten results—Poe's "The Fall of the House of Usher" is nowhere to be found.

With its shortcuts Yahoo makes no pretense of objectivity—it is clearly steering searchers toward its own editorial services, which it believes can satisfy the intent of the search. In effect, Yahoo is saying "You're looking for stuff on Usher? We got stuff on Usher, and it's good stuff. Try what we suggest; we think it'll be worth your time."

Apparent in that sentiment lies a key distinction between Google and Yahoo. Yahoo is far more willing to have overt editorial and commercial agendas, and to let humans intervene in search results so as to create media that supports those agendas. Google, on the other hand, is repelled by the idea of becoming a content- or editorially driven company. While both companies can ostensibly lay claim to the mission of "organizing the world's information and making it accessible" (though only Google actually claims that line as its mission), they approach the task with vastly different stances. Google sees the problem as one that can be solved mainly through technology—clever algorithms and sheer computational horsepower will prevail. Humans enter the search picture only when algorithms fail—and then only grudgingly.

But Yahoo has always viewed the problem as one where human beings, with all their biases and brilliance, are integral to the solution. It's humans, backed by technology, who drive the "also try" results at the top of the page (the process has been automated, but it is classic architecture of participation stuff: "here's what other human beings find useful related to your search"). It's humans, backed by technology, who push Yahoo's internal content and commerce sites to the fore in the shortcut results. DNA has much to do with it: Yahoo started as an entirely subjective collection of links (Jerry's and David's Guide to the World Wide Web), and the first few years at Yahoo were dominated by its human-edited directory. Humans first, technology second.

Google, on the other hand, started as an extremely clever algorithm that solved an intractable and recursive mathematical problem. Technology first, humans second. Over the past four years, Google has changed on this front—if you asked anyone there in 2002 whether it was a media or a technology company, the answer was always technology. Ask now, and it depends on whom you ask. But the furthest even the most media-savvy person within Google will go is to say, "We're a media-driven technology company." At Yahoo, everyone there understands it is a major player in the media business, from Terry Semel down.

As both companies move forward with new features and services, I expect this distinction will surface in any number of interesting and important ways. Both approaches have their merit; both have succeeded and will continue to do so. But expect some tension over the next few years, in particular with regard to content. In late 2004, for example, Google announced it would be incorporating millions of library texts into its index, but made no statements about the role the company might play in selling those texts. A month later, Google launched a video search service, but again stayed mum on if and how it might participate in the sale of television shows and movies over the Internet. (That might be changing. In June 2005, the *Wall Street Journal* reported that Google was close to launching a payment system similar in scope to eBay's PayPal.)

Google is clearly in the process of declaring its position relative to the content industry, and it seems to be this: we will become your distribution sugar daddy. We'll be Switzerland—allow us to index your content, and when people find it through us, we'll enable you to sell it. This approach became more apparent with the discussion and disclosure of a 2004 patent application in Google's name that creates a system by which media is discovered and then paid for.

In such a system, one can imagine that Google has or will cut deals with any number of content owners and somehow incorporate that content into its index (the company has been rumored to be doing just that, but refuses to comment). When you search for something, let's say "usher," the actual content that Usher has created will come up in the results, and thanks to the distribution deals Google has cut, you can buy that content right there on the spot. Everyone gets paid!

With Yahoo, of course, this already happens. But for Google to put itself into the position of media middleman is a perilous gambit—in particular given that its corporate DNA eschews the almighty dollar as an arbiter of which content might rise to the top of the heap for a particular search. Playing middleman means that in the context of someone looking for a movie, Google will determine the most relevant result for terms such as "slapstick comedy" or "romantic musical" or "Jackie Chan film." For music, it means Google will determine what comes first for "usher," but it also means Google will have to determine what should come first when someone is looking for "hip-hop." Who gets to be first in such a system? Who gets the traffic, the business, the profits? How do you determine, of all the possibilities, who wins and who loses?

In the physical world, the answer is clear: whoever pays the most gets the positioning, whether it's on the supermarket shelf or the bin end of a record store. As Yahoo also becomes a superdistributor of media content, I have no doubt the company will figure out some way to index and distribute media content that is moderated by the traditional market forces of who pays the most, and what is the most popular.

But Google, more likely than not, will attempt to come up with a clever technological solution that attempts to determine the most "objective" answer for any given term, be it "romantic comedy" or "hip-hop." Perhaps the ranking will be based on some mix of PageRank, downloading statistics, and Lord knows what else, but one thing is certain: Google will never tell anyone how it came to the results it serves up. Which creates something of a catch-22 when it comes to making money. Will Hollywood really be willing to trust Google to distribute and sell its content absent the commercial world's true ranking methodology: cold, hard cash?

In the end, both companies are in the same business, and were I forced to name that business in one word, I'd argue that it is media.

Yes, Google started its life as an algorithm in a PhD program, and Yahoo started as an edited guide to the Web, but they are clearly converging into the same space; they mediate information and services for consumers, and derive value from those services using the traditional revenue streams of the media business—advertising and subscriptions. (Google may not play in the subscription business yet, but I'd wager it will, and shortly. I asked Brin about this and he answered that he could imagine a day when Google would begin taking referral fees, at a minimum.)

Because of its media DNA, Yahoo is clearly more comfortable with extracting fair value for media services rendered, and because of that, I believe it has been free to innovate in its approach to search: as one of Yahoo's executives recently put it to me, "We are entirely focused on completing tasks." In other words, if the task at hand is buying an Usher CD, or checking a flight, or finding a local restaurant, Yahoo has repeatedly innovated in building a suite of search results that help a consumer complete the task and get Yahoo paid in the process.

When it comes to completing tasks, Google does the same in many instances, but the company has been uncomfortable with the idea of tying commerce to its media products—it resists making money on the value created in any way other than by AdWords (and even resisted that, at first). Two examples are News, where there is no business model, and Froogle, where the only model is AdWords. In a way, this reluctance gates innovation in the search results space. If the consumer truly wants to shop, or browse high-quality news results, and you provide a great service to do so, there's no shame in making a buck while doing it, even if that buck is made in ways other than by advertising (such as cutting deals with music or news publishers, or selling your consumers up to a premium service if you can).

Certainly, Google is a major media player. And the cards it holds, combined with the moves it has made recently, point toward its being an even larger force in media in the future. A case in point is Google Print. As that program expands, a number of questions arise. How will Google make money from books whose copyright has expired? As it brings hundreds of thousands of out-of-print books onto the Web and into its index, will it allow others to access and index that new treasure trove, or will it act more like a traditional media company, which would "own" that resource for itself? How will it choose what it brings into the index? Will it start with those items that might sell the best or those it considers in some way "good for the world"? With regard to books that are in print, will it limit itself to being solely an organizational tool supported by AdWords, or might it start to take a percentage of sales for books that are sold via the Google Print service? And will the print model scale to television, movies, or music?

Pure organic search made Google what it is, and remains the true north of the company. At Yahoo, pure organic search is viewed as one (extremely important) option among a range of search-related services that the company provides. When you enter a search term, pure organic results are always there, but so are other services that the company has developed in response to the implied intent of your declarative term. In early 2005, America Online, a Google partner, announced a new search strategy that aligned itself with Yahoo's approach. Not surprisingly, AOL is owned by Time Warner, a media company.[4]

When companies like Google and Yahoo become mediators of content such as books and videos, what becomes of companies like Amazon? Think about that one a bit, and it becomes much more obvious why Amazon is busy perfecting A9.com, its own search engine. Search drives commerce, and commerce drives search. The two ends are meeting, inexorably, in the middle, and every major Internet player, from eBay to Microsoft, wants in. Google may be tops in search for now, but in time, being tops in search will certainly not be enough.

Google understands this. As this book was going to press, it announced a new portal strategy called Fusion, which allows Google users to customize their home pages and intergrate all of Google's services into one view—just like AltaVista, Excite, Yahoo, AOL, and MSN.

Inventing the Future

By 2005, Google was adding nearly four employees a day. In an article in the *New York Times* in February, Sergey Brin announced an innovative incentive program for his growing number of employees. Labeled the Founders' Awards, the new program promised millions of dollars in compensation to employees or teams that the founders felt significantly increased Google's overall value. "Periodically we buy little companies that have accomplished a great amount," Brin told me, explaining the new program. "We might buy them for ten million dollars or something. If [we didn't have the Founders' Awards], then I'm basically telling people 'Don't join Google. Go to a little start-up and then get acquired by Google.' "

Brin's program is an acknowledgment of the reality that hits every successful technology start-up headed into middle age—the market tends to reward maverick innovation outside of the mainstream. In the late 1980s and early 1990s, another technology giant—Microsoft—had this same problem. Scores of its most innovative employees left the company to start businesses, many with the idea of simply selling their company back to Microsoft once the time was right. Brin aims to nip that trend in the bud. "Ultimately, I believe that everybody should have the opportunity to make ten million dollars," he told me. Who wouldn't want a boss like that?

But Google will face more than competition and the looming issue of retaining its best and brightest. Its most important job will be to continue to innovate. I'd add one detail to that—the ability to innovate in a focused, market-driven fashion. Despite its reorganization, Google remains an extraordinary experiment in bottom-up innovation. According to Brin, the company still does not plan what new products or markets to enter—the ideas still come from the ranks of its employees, as opposed to any centralized planning process. (Such a claim is hard to take entirely seriously, especially when it comes to Google's advertising products. In April 2005, for example, Google announced it was revamping its AdWords product to accept image advertising based on a CPM model—the very kind of ad Brin and Page dismissed earlier in Google's history. This move was clearly strategic in nature and not the result of any bottom-up engineering innovation. Google realized how large the advertising business is beyond paid search, and it moved accordingly.)

Given that Google has what is widely considered to be the most extensive computing platform on the face of the planet and an extremely talented workforce, the company clearly has a good base on which to build. But what might it do with that platform?

Speculation on Google's next move is a full-time occupation for hundreds of analysts in the Valley and on Wall Street, and the company's every fidget can impact vast ecologies in the media, commerce, and knowledge industries. It's best to start with posing that question to the company's leaders, then work out from there. Of the three, only Schmidt is willing to speculate in any meaningful fashion.

"Google hopes to help you find anything," Schmidt told me. "We need to keep inventing new ways of using our data centers and the information we have assembled. Google has one of the largest data centers in the world, and one of the largest collections of bandwidth in the world. What are the technological possibilities of that platform? We have conversations about how you take the many tens of thousands of computers we have, and build platforms that enable people to do things at a scale that was not previously possible in the world."

By all means, do tell, I urged him. What might you build next? "We understand that video is the next holy grail," Schmidt replied. "How many camcorder tapes do you have?"

I answered that I had no idea, but a lot, at least a boxful. "If the average reasonably high-income person had a hundred each, that's millions and millions of tapes," Schmidt said. "That certainly sounds like an unsolved problem."

So is that it? The future of Google is—indexing your old video collection? Somehow, I figured Schimdt was being a bit disingenuous. Certainly helping people digitize, index, organize, and access their personal information, whether it is in e-mail, videos, photographs, or documents, is in Google's future. The company already has several products (Picasa, Google Desktop) that address many of those needs. And making personal media accessible is a huge accomplishment in itself. But it doesn't feel—well—*big enough* for the likes of Google. I pressed Schmidt—what are some of the really cool projects on that list of five hundred or so Brin was reviewing in his shaking chair?

"We do have an 'other' category," Schimdt said, referring to the six categories of Google's new corporate structure. "The joke there, of course, is that the carbon fiber nanotubes to the moon go in that category."

OK. But when you get Schmidt to focus on the more immediate and plausible future, the furthest he'll go is to lay out a scenario where Google's core business model—AdWords—is extended to its most far-reaching potential. In early 2005, Google rolled out a service that gave advertisers far more control of their AdWords programs. Using this tool, a business could theoretically manage thousands, if not millions, of keywords—as many keywords as there might be things to sell. As Schmidt told *Fortune* magazine: "Pick any large consumer packaged goods company. How many products do you think they have? Probably millions, I would think, by the time you have all the variants and the different geographies and legal rules. We want every one of those products to be advertised in the appropriate market within Google in the right country. That's our goal."

If you add in every small business in the world—and believe me, Google is thinking that way—you can sum up Google's ambitions in the commercial world as this: the company would like to provide a platform that mediates supply and demand for pretty much the entire world economy. As Schmidt put it, "The sum of

[Google's addressable] market, if you include the large companies and the small companies throughout the world, is the world's gross domestic product."

"We think of it as a marketplace," Schimdt added.

In other words, the market for Google's core business—or Yahoo's or Microsoft's, not to mention eBay's and Amazon's—has hardly been scratched. Even more fascinating, as more and more buyers and sellers come online, searching either for customers or for products, Google's AdWords morphs from an advertising play into something more like eBay's model. It's no coincidence that eBay is the one company whose margins and revenues are growing as quickly as Google's. In a perfect market, where demand is simply one computable bit of information, and supply another, matching the two is an extremely lucrative business.

So Google is angling to become the de facto marketplace for all of global commerce, unseating eBay in the process. OK, that's big, but is it big enough to fulfill the world's expectations for this company? When you poll folks outside of Google who are nevertheless extremely smart on the company's intentions, and you listen very, very carefully to the public pronouncements of its senior engineers and leaders, a reasonably clear picture begins to emerge of a future for the company that is even larger.

When grasping for precedents that might explain this future, only one will suffice: Microsoft. Over the course of three decades, Microsoft became one of the most valuable companies in the world by relentlessly focusing on its core mission of a computer on every desk, and Microsoft products running on every computer. Audacious as this goal was when stated by founder Bill Gates back in the late 1970s, Microsoft pretty much achieved it, in the developed world anyway, within twenty years.

Now let's parse Google's audacious goal: to organize the world's information and make it accessible. Note that the word "search" is not in the mission—search is, in the end, the presumption, one side of an equation that presumes something needs to be found. And how might anything—recall Schmidt's words, *Google wants to help you find anything*—be found?

The answer is simple: forget about a computer on every desk. Instead, the entire world needs to become computerized. And to many observers of Google's strategy, that's exactly what the company is out to take advantage of.

Let's break down Google's mission even further. What is "information," anyway? In the end, it's data that describes something, anything. Maybe it's a document on the Web, but to think that's where it ends is to think small. Perhaps it's the location of your GPS-enabled keys, or the cost of a box of Pampers on a store shelf in suburban Miami. It could be your wedding photos, or a real-time video stream of a tsunami racing across the Indian Ocean. If the first few years of Google's rise to dominance have taught us anything, it is this: if something is of value, it needs to be in Google's index. What happens if the entire world becomes the index?

Thinking about the merger of the physical world with the World Wide Web might make your head hurt, but after you've reached for the aspirin, Google's mission starts to resonate with slightly larger ambitions. Information is all around us, but how might the company make it accessible?

This is where the concept of a Web operating system comes in. Recall Microsoft's success in driving a computer to every desk, with Windows on every computer. The next step in the evolution of the computer was clearly the connection of every computer to every other—what came to be known as the Internet. But what's next after that?

According to many leading-edge computer scientists and theorists, the Web is in the process of becoming the next great computing platform—the successor to Microsoft Windows, owned by no one but used by everyone. And the Web is also in the process of connecting to everything—be it a desktop computer, a mobile phone, an automobile, or a set of keys. Given that, the theory goes, the companies best positioned to deliver hugely scaled services over the Web platform are best positioned to win. And when it comes to hugely scaled services, nothing beats search.[5]

Google's mission of organizing the world's information and making it accessible sets the company up to deliver nothing short of every possible service that might live on top of a computing platform—from mundane applications like word processing and spreadsheets (Microsoft's current bread and butter) to more futuristic services like video on demand, personal media storage, or distance learning. Many experts believe that in the near future, we'll store just about everything that can be digitized—our music, photographs, work documents, videos, and mail—on one massive platform: the Google grid.

In other words, Google has, in its seven short years of corporate life, become a canvas upon which we project every application or service that we can imagine might arise in our increasingly digital future. Google as phone company? As cable provider? As university? As eBay, Amazon, Microsoft, Expedia, and Yahoo all rolled into one? It's conceivable; and that, in the end, is what makes the company—and search, the application that spawned it—so fascinating to us all. Nothing beguiles like the promise of unlimited potential. For now, anyway, Google holds that promise.

At the end of a long conversation about her company that touched on this point, I asked Susan Wojcicki, one of Google's early senior managers, if she ever thinks about such things, whether the weight of the world's expectations ever gets too heavy to hold.

"Sometimes I feel like I am on a bridge, twenty thousand feet up in the air," Wojciki replied with an inward gaze. "If I look down, I am afraid I'll fall. I don't feel like I can think about all the implications."

Chapter 11
Perfect Search

There will always be plenty of things to compute in the detailed affairs of millions of people doing complicated things.
—"As We May Think" by Vannevar Bush

All collected data had come to a final end. Nothing was left to be collected.

But all collected data had yet to be completely correlated and put together in all possible relationships.

A timeless interval was spent in doing that.

And it came to pass that AC learned how to reverse the direction of entropy.
—"The Last Question" by Isaac Asimov

Where do we go from here? Now that Google is public, and revealed to be mortal, now that almost every major media and information technology company in the world has declared search integral to its future, what might come next? Can anything possibly match the cultural thunderclap of the early Web, or the singular epiphany we all felt the first time we used Google?

Of course it can. When it comes to search, as with the Internet itself, the most interesting stuff is yet to come. As every engineer in the search field loves to tell you, search is at best 5 percent solved—we're not even into the double digits of its potential. And search itself is changing at such a rapid pace—in the past year important innovations have rolled out once a week, if not faster—that attempts to predict the near future are almost certainly doomed.

So let's instead imagine a world of perfect search. What might that look like? Imagine the ability to ask any question and get not just an accurate answer, but *your* perfect answer—an answer that suits the context and intent of your question, an answer that with eerie precision is informed by who you are and why you're asking. This answer is capable of incorporating all the world's searchable knowledge into the task at hand—be it captured in text, video, or audio formats. It's capable of distinguishing between straightforward requests (Who was the third president of the United States?) and more nuanced ones (Under what circumstances did the third president of the United States forswear his views on slavery?).

While it's true that most questions don't have an objectively perfect answer, perfect search would provide *your* perfect answer, as you determine it—in a report form, perhaps, or by summarizing key points of view and trends. This perfect search also has perfect recall. It knows what you've seen and can distinguish between a journey of discovery, where you want to find something new, and recovery, when you want to find something you've seen before. And, quite important, it's capable of distinguishing between a document and a person—and suggesting that to get the perfect answer, you may well best talk to *this person,* as opposed to reading *that document.*

In short, the search engine of the future isn't really a search engine as we know it. It's more like an intelligent agent—or as Larry Page told me, a reference librarian with complete mastery of the entire corpus of human knowledge.

That's a long way from the typical search engine of today, but imagining such a service no longer falls into the realm of science fiction. It's the stated goal of nearly major player in search, be it IBM, Microsoft, Google, Yahoo, or scores of others.

But how do we get there, and if we do, how might that change the world? Such an engine would require that we solve scores of ridiculously difficult computer-science problems. Let's look at a number of them in turn.

Search Everywhere

First, let's say this clearly: in the near future, search will metastasize from its origins on the PC-centric Web and be let loose on all manner of devices. This has already begun with mobile phones and PDAs; expect it to continue, viruslike, until search is built into every digital device touching our lives. The telephone, the automobile, the television, the stereo, the lowliest object with a chip and the ability to connect—all will incorporate network-aware search.

This is no fantasy; this is simple logic. As more and more of our lives become connected, digitized, and computed, we will need navigation and context interfaces to cope. What is TiVo, after all, but a search interface for television? ITunes? Search for music. That box of photographs under your bed and the pile of CDs teetering next to your stereo? Analog artifacts, awaiting their digital rebirth. How might you find that photo of you and your lover on the beach in Greece from fifteen years ago? Either you scan it in, or you lose it to the moldering embrace of analog obscurity. But your children will have no such problems; their photographs are already entirely digital and searchable—complete with metadata tagged right in (date, time, and soon, context).[1]

But let's not stop our digital fantasy train yet. It may sound farfetched, but in the future, your luggage will be searchable. Within two decades, nearly everything of value to someone will be tagged with tiny computing devices, devices capable of saying, upon radiowave-based query, "I'm here, *right here,* and here's what I've been doing while you were away." Instead of the ubiquitous bar codes airport officials now slap onto your luggage, there'll simply be an RFID (radio frequency ID) chip. Lost your luggage? I don't think so. Not when you can Google your Louis Vuitton in real time.

Think about that—Google your dog, your kid, your purse, your cell phone, your car. (Do you have an E-ZPass or OnStar yet? You will.) The list quickly stretches toward the infinite. Anywhere there might be a chip, there can and most likely will be search. But for perfect search to happen, search needs to be everywhere, attached to everything.

This means that among many other things, search needs to solve what so far has been a rather intractable problem: that of the invisible Web. As Gary Price and Chris Sherman point out in their book of that name,[2] the invisible Web comprises everything that is available via the Web, but has yet to be found by search engines. Deep databases of knowledge, like the University of California's library system or the LexisNexis news and legal citation service, are walled off from search for commercial or technological reasons. And while the contents of your hard drive may be digital, they most likely have not been indexed and offered up to a search application—yet. As I pointed out earlier, all the major search engines have launched desktop search tools which index your hard drive and serve up the results much as you might see Web results. Prior to the advent of desktop search, your PC was part of the invisible Web. No longer.

Also mostly invisible, and mainly still stuck in the analog world, is what might be called the content Web. There are nearly 100 million books extant, but only several hundred thousand online as of this writing. Add to that unsearchable pile humanity's analog archives of film, television, and periodicals.

Thanks to Napster, we've already got the music nut partially cracked. When Napster launched, millions of people ripped copies of their favorite music to the Web. And therein most likely lies the solution to the rest of our previously unsearchable media. For nearly every book, film, and television show, someone, somewhere, will come up with a reason to put it on the Web, assuming we can get out of our own way with regard to intellectual property issues.[3] Massive archiving projects, such as Google Print, the Internet Archive, and Amazon's Search Inside the Book, have gone a long way toward solving a piece of this problem, but they have a long, long way to go, and simple logic tells us that no one entity can (or should) archive the sum total of humankind's information. No, when it comes to making the world searchable, the best way is to simply let the world do it.

This phenomenon has many casual monikers, but I like to call it the Force of the Many. Eventually, everything of value—including your luggage—will be connected to the Web, because to be connected is definitional to the concept of value in a wired world. As the Force of the Many weaves humanity's belongings into the Web, search engines will weave this new content into their indexes, moving the world ever closer to the possibility of perfect search.

The Clickstream

Ubiquity is critical to perfect search, but it means nothing if the engine does not understand *you*—your likes and dislikes, your tendencies and tics. How might an engine be not only ubiquitous but also personal?

A solution to this problem lies in the domain of your clickstream. Through the actions we take in the digital world, we leave traces of our intent, and the more those traces become trails, the more strongly an engine might infer our intent given any particular query. Many services have begun tracking our trails, and over time I expect those trails—the sum total of which makes up the Database of Intentions I discussed in Chapter 1—to turn into relevance gold.

A clickstream might best be understood as a story by another name. We love stories—they are how we understand the world. Were I to tell a friend what happened at last night's ball game, I wouldn't send him a box score. I'd say something like "We looked terrible in the first two innings. Our rookie pitcher was tight and we had back-to-back errors resulting in a three-run deficit by the second. But then Snow nailed a three-run homer that put us back in the game, and in the fifth we rang up three more. It was all Giants from then on!" A story is our way of taking a journey and making it portable so we can share it with others.

So here's a story about one clickstream. In the summer of 2004, I was researching the phrase "tempting fate" for Chapter 9. I had a hunch the phrase would relate to Google's IPO and its engineering-driven culture. I was sure that the phrase originated in Greek or Roman mythology—proof that human beings have always struggled with the questions of determinism, the gods, free will, and destiny. (The story of Odysseus lashing himself to the mast of his ship so as to hear the song of the Sirens came to mind. But while that was tempting *his* fate, Homer postdated most Greek myth.) At the very least, there had to be a good story behind "tempting fate," right?

So what did I do? I fired up Google and started poking around. I started with the simple query "tempting fate," but the results were far too broad (though it was interesting to see a Google News story about the Athens Olympics). I called my mother, a middle-school English teacher with knowledge far superior to my own when it comes to mythology, and she reminded me that Shakespeare often used the Fates in his work. Armed with this new high-order bit, I went back and Googled "the Fates mythology."

I was onto something. I found a site that chronicled the three Greek deities of fate, and using information from there, I Googled my way through all manner of references to the Fates. But I couldn't find the perfect answer: who first tempted fate? Perhaps someone famously coined the phrase, I thought. Or perhaps there was no perfect founding mythology. I suddenly got an odd sense of déjà vu—I remembered that I had seen a site a few weeks back that would be very useful to my current search. In an earlier search session, I had come across a great resource for quotations and literary references.

But alas, I did not save the URL. If I had access to that prior clickstream—my search history—I could have quickly found it. But instead, I had to start all over again.[4]

While I never did find that quotation site, I did find myself on a great journey, from early twentieth-century texts on philosophy and religion to scholarly interpretations of the Fates and their role in early Greek tragedy. Along the way, I got to brush up on Homeric epics, Shakespeare, Joyce—it was great fun. And in the end I came to a much fuller understanding of my original question, which was this: why on earth would Google launch the bidding process for its shares on Friday the thirteenth? *Why tempt fate?*

I found my own highly subjective version of the answer. As I said in Chapter 9, it was that engineers, like Greek philosophers, believe that fate can't be tempted—but I didn't come to that conclusion by clicking on one of the first ten results of my initial Google search. I found it by going on a journey, one that now, through the telling, you've all gone on as well.

But what may well become possible in the world of perfect search is the ability to take the clickstream of that journey and turn it into an object—a narrative thread of sorts, something I can hold and keep and refer to, a prop to aid in the telling and retelling of how I came to my answer. Tracks in the dust, so to speak, that others can follow, or question to discover how I came to my conclusions. And these tracks are not just potential narratives for others to read; they can also be objects that can be spidered by a search engine, providing them with an entirely new order of intelligence about how people learn. In the aggregate, these clickstreams can provide a level of intelligence about how people use the Web that will be on an order of magnitude more nuanced than mere links, which formed the basis for Google's PageRank revolution.

"As We May Think," Vannevar Bush's famous 1945 essay in *The Atlantic,* posited the memex, a computational machine that created the equivalent of clickstreams in the field of scholarly research. In the essay, Bush outlined a looming problem for humankind—that knowledge and learning have become so complicated, so layered, so inefficient, that it is nearly impossible for anyone to be a generalist, in the sense that Aristotle was in his day. In short, there is simply too much knowledge—we can't depend on any one person to be a philosopher to the kings.

As Bush outlined it, the memex gains its potency by capturing the traces of a researcher's discovery through a corpus of knowledge, then storing those traces as intelligence so the next researcher can learn from and build upon them.

Clickstreams are the seeds that will grow into our culture's own memex—a new ecology of potential knowledge—and search will be the spade that turns the Internet's soil. Engines that leverage clickstreams will make link analysis–based search (nearly all of commercial search today) look like something out of the Precambrian era. The first fish with feet are all around us—nearly every search engine now supports search history, and dozens of interesting tools have recently come to market that attempt to make sense of the patterns we searchers are leaving upon the Internet's corpus. We have yet to aggregate the critical mass of clickstreams upon which a next-generation engine might be built, and it will not necessarily be built with our tacit consent, as I pointed out in Chapter 8. But regardless of our trepidation, we're already pouring its foundations.

Local and Personal

But while such third-generation search engines have yet to appear, what *is* here, at least in its first phase, is personalized search, specifically the particular variant known as local search. The idea behind personal search is pretty simple: the more an engine knows about you, the more it can weed out irrelevant results. Ask.com, Google, Microsoft, and Yahoo have rolled out some flavor of personalized search in the past few years, and most experts predict big things for this feature in the future.

As with nearly everything, Google and Yahoo take entirely different tacks in their initial approaches to the personalization problem. Google has yet to fully integrate personalization into its main index, but it does integrate local searches. Google's version of local has two inputs: the search term itself and a bit of local information (such as a zip code or town name). It then folds Google search results into yellow pages results.

It's very much in Google's character not to assume too much about the person typing queries into the search box, but Yahoo does it as a matter of course. If you type "giants scores" into Yahoo, you'll get a box score of the game in process as the top result.

The term for what Yahoo is doing when I type in "giants scores" is "inference"—Yahoo has programmed the engine to infer what I intended, and to present results that more likely than not will be extremely relevant. (Yahoo calls this feature shortcuts; and AOL, which introduced similar technology in early 2005, calls it programmatic search.) Yahoo, AOL, Ask, and others do this for movie listings, music, and other obvious topics, but the real question is whether this approach can scale to less obvious topics.

Yahoo Local is another example of this approach. Instead of simply providing you with localized Web results based on a zip code and yellow pages, Yahoo finds new ways to surface, sort, and present information that attempts to understand the intent of your query. The service invites you to navigate your way toward your perfect answer, a process I believe we'll see far more of in the future. Search scholar and entrepreneur Ramesh Jain has called this approach giving search a steering wheel—a control mechanism for driving through your search results.[5]

The use of search as an interface steering wheel got a boost when Yahoo introduced Y!Q, a contexual search-anywhere technology. Y!Q could potentially shift the way that consumers access and interact with search technology. "With the introduction of features like shortcuts, we have broken through one of the oldest linear search paradigms: input query, review results, input query, review results, et cetera," says Yahoo SVP/Search Jeff Weiner. "Our goal with Y!Q is to increase the accessibility to search when and where users are most likely to be inspired to conduct a search," Weiner told me.

In other words, search will happen anywhere on the Web, not just at a destination site like Google or Yahoo Search. To this end, in early 2005, Google introduced the Google Deskbar, a floating search box that lives anywhere on your desktop, and a set of applications programming interfaces (APIs) that allow any desktop software supplier (like Adobe, maker of the popular Photoshop application, for example) to plug into Google's infrastructure.

As your desktop becomes more integrated with search, your results won't simply be a list of URLs but an on-the-fly report about the topic you've indicated, delivered instantly to you wherever you happen to be—whether it's in an Excel spreadsheet or out on the Web. If, for example, you're reading a news story about a new band, and you want more information about the band, you can click on a Y!Q icon and instantly the search service will access a discography, as well as offer you reviews, music videos, or the ability to purchase an album.

And this approach to search need not be limited to popular queries with obvious structural results (like bands or movie listings). In the future, this kind of a search shortcut could deliver results on any query you might have, tailored to who you are, what you are reading, and your past search history. If I had this kind of search technology at my disposal while looking up "tempting fate," for example, I might have had my answer in an instant.

For another compelling view of this personalized future, head to A9.com. But be prepared to use it for a while, as its most interesting features don't kick in until you've logged some time and built your own search history.

Udi Manber, A9's CEO, has spent the past fifteen years of his life thinking about search, and when he left Yahoo in 2003 to run A9, it was major news in the search community. The first fruits of his new company's efforts debuted in spring of 2004. The engine employed Google's index of Web sites, but layered a robust interface on top and integrated Amazon's Search Inside the Book feature, which shows you a full page of a book's text surrounding any keyword or phrase you are searching for. A9 was also the first engine to employ the concept of search history in its results (Google has since introduced it as well). If you install A9's toolbar software, it will even remember where you've been on the Web as well—your complete clickstream. Coupled with a number of other innovative features, A9 was a clear declaration by Amazon that it was a significant search player, one to watch as the ongoing push-pull drama between search and commerce unfolds.

Search as the New Interface

Jain's steering wheel metaphor resonates because he views search as an interface—a way to navigate in our increasingly complicated computing environment. Search as most of us know it has for years been stuck in what Tim Bray, a search pioneer now at Sun Microsystems, calls the C-prompt phase. Like DOS before Windows or the Macintosh, search's user interface is pretty much command driven: you punch in a query, you get a list of results. Many companies have attempted to address this shortcoming, but until recently they lacked a key element necessary to truly make an interface breakthrough in search.

That key element is your clickstream. Given that nearly every major search engine has a search-history feature, it won't be long before we begin to see significant changes in how results are tendered to us. By tracking not only what searches you do, but also what sites you visit, the engines of the future will be able to build a real-time profile of your interests from your past Web use. They can then fold that profile into both your search results and the search interface itself, making for what can become, with regular use, an entirely new approach to searching. Call it searching your personal Web—search enhanced by everything you've seen, every query you've clicked on, and every page you've bookmarked or otherwise interacted with.

On A9.com, you can view search results as more than just a list of URLs. Instead, you can see various "panes" of result information—images, for example, or your history, or results from partner sites rich in structured information (such as dictionaries, medical sites, or the Internet Movie Database). The more you interact with this interface, the richer it becomes.

True to Jeff Bezos's observation, A9 has broken search into its two most basic parts. Recovery is everywhere you've been before (and might want to go again); discovery is everything you may wish to find, but have yet to encounter. A9 attacks recovery through its search history feature and its toolbar, which tracks every site you visit. The discovery feature finds sites you might be interested in on the basis of your clickstream and—here's the neat part—the clickstream of others.

This powerful feature smells an awful lot like Amazon's fabled recommendation system and, over time, may well become the basis of an entirely new relevance scheme that builds upon Google's link-based PageRank. A9 is something of a Web information management interface, with search as its principal navigational tool.

Through innovations like Google Deskbar, A9, and Y!Q, the search interface will evolve well beyond what we see today. Search will swallow untold petabytes of previously unindexed data—from media like books and films to reference databases like GuruNet and LexisNexis, to objects like luggage and bottles of wine, to your own personal Web through desktop search and search history. And those same engines will then parse all that data not just with the blunt instrument of a PageRank-like algorithm, but with subtle and sophisticated calculations based on your own clickstream and those of millions of others. The result? Yet another step toward finding the perfect answer to your search.

The Semantic Web?

But perfect search will require more than ubiquity, clickstreams, and personalization. The vast corpus of information now available to us is often meaningless unless it is somehow tagged—identified in such a way that search engines can best make sense of it and serve it up to us. Many in the search industry believe search will be revolutionized by what is called metadata. Clickstreams are a form of metadata—information about where you go and what you choose as you browse the Web. But to get to more perfect search, we need to create a more intelligent Web. That means tagging the relatively dumb Web pages that make up most of the Web as we know it today with some kind of code that declares, in a machine-readable universal lingo, what they are, what they are capable of doing, and how they might change over time.

This is the vision of the semantic Web, as it is known by those responsible for its conception and furtherance. It remains—for the most part—an unrealized but a rather compelling dream. None other than Tim Berners-Lee, father of the Web, is its main proponent. Way back in 1998, Berners-Lee's "Semantic Web Road Map" outlined a universal and relatively simple approach to structuring metadata so that the Web becomes more intelligent. While it's always dangerous to lean too heavily on metaphor, the basic idea is that with semantic tags, the Web becomes more like a structured database such as Lexis-Nexis or the Sabre reservation system, making it far easier to find things. This in turn allows rules of logic, or reason, into the equation.

This structure also makes it much easier to *do* things, to execute complex tasks built upon finding things—scheduling a meeting, planning a trip, organizing a wedding, you name it. In a seminal *Scientific American* article in May 2001, Berners-Lee and his colleagues explained:

The real power of the Semantic Web will be realized when people create many programs that collect Web content from diverse sources, process the information and exchange the results with other programs. The effectiveness of such software agents will increase exponentially as more machine-readable Web content and automated services (including other agents) become available. The Semantic Web promotes this synergy: even agents that were not expressly designed to work together can transfer data among themselves when the data come with semantics.

In another paper, Berners-Lee goes on to explain the impact this might have on search:

If an engine of the future combines a reasoning engine with a search engine, it may be able to get the best of both worlds. . . . It will be able to reach out to indexes which contain very complete lists of all occurrences of a given term, and then use logic to weed out all but those which can be of use in solving the given problem. . . .

I also expect a strong commercial incentive to develop engines and algorithms which will efficiently tackle specific types of problem. . . .

Though there will still not be a machine which can guarantee to answer arbitrary questions, the power to answer real questions which are the stuff of our daily lives and especially of commerce may be quite remarkable.

Berners-Lee's vision of a semantic Web may be a long way off, but there are thousands of alpha geeks working on pieces of it, and its core coding language, called resource description framework (RDF), has become a standard among most cutting-edge Web technologists. In 2002, Paul Ford, an author and leading semantic Web thinker, wrote a piece that tied Berners-Lee's ideas to the reality of the then-emergent power of Google. Entitled "August 2009: How Google Beat Amazon and eBay to the Semantic Web," the essay began as a primer on RDF, but quickly grew into one of the Internet industry's favorite Google scenarios.

To quote the essay:

Enter Google. By 2002, it was the *search engine, and its ad sales were picking up. At the same time, the concept of the "Semantic Web," which had been around since 1998 or so, was gaining a little traction, and the attention of an increasing circle of people.*

So what's the Semantic Web? At its heart, it's just a way to describe things in a way that a computer can "understand." Of course, what's going on is not understanding, but logic, like you learn in high school:

If A is a friend of B, then B is a friend of A.

Jim has a friend named Paul.

Therefore, Paul has a friend named Jim.

Using a markup language called RDF . . . you could put logical statements like these on the Internet, "spiders" could collect them, and the statements could be searched, analyzed, and processed. What makes this different than regular search is that the statements can be combined. So if I find a statement on Jim's web site that says "Jim is a friend of Paul" and someone does a search for Paul's friends, even if Paul's web site doesn't have a mention of Jim on it, we know *Jim considers himself a friend of Paul.*[6]

But Ford didn't stop there; he took it a step further and showed how, once the semantic Web took root, Google might become a global marketplace far exceeding even eBay or Amazon. In essence, once you have good information about things for sale, and good search connecting them, it's relatively trivial to be in the business of putting the two together.

But a major hurdle to the rise of the semantic Web has been standards: who gets to say which tags are right for which pages? If there is a picture of a Cape Cod seashore on the Web, should it be tagged as "beach," "shore," "ocean," or any number of other possible words? As Yahoo learned early in its directory days, the nearly limitless possibilities of the Web do not lend themselves to top-down, human-driven solutions.

Again, this is where the Force of the Many comes in. In late 2004 and throughout 2005, a new kind of tagging scheme arose, one based not on any strict, top-down hierarchy, but rather on a messy, bottom-up approach. Small start-up companies like Flickr, Technorati (a weblog search engine), and del.icio.us (a link-sharing site) began giving their users the ability to tag anything they saw, and then to share those tags with others. By letting anyone tag anything, the theory goes, ultimately a kind of fuzzy relevance for any given item will emerge. The photo of the Cape Cod seascape, for example, will probably be tagged with all the possible descriptors. That way, no matter what phrase a person uses to search for it, whether it's "ocean photos" or "Cape Cod seascapes," that photo will be found.

Early bloggers dubbed this approach folksonomies—folk + taxonomy—and the movement is gaining momentum. Yahoo's purchase of Flickr for an estimated $15 million to $30 million gave tagging an early boost. Flickr had no revenue, so clearly Yahoo saw value somewhere else. Given how important search is to Yahoo, it's a fair bet that Yahoo saw value in Flickr's tagging scheme.

What Have Blogs Got to Do with It?

Yet another development related to the semantic Web is the recent explosion of blogs and syndicated feeds (often referred to as RSS, for real simple syndication). At this writing, there are 8 million to 12 million active blogs on the Internet, and millions more RSS feeds, which are simply "portable" versions of blogs or other media sites that can be read via applications called newsreaders.

Blogs are home pages of sorts, but they are far more than that—they represent a new form of authoring on the Web, authoring that takes as its foundation the ability to quickly and easily link to anything else on the Web. Back when PageRank was born and Web pages were hand-rolled using laborious HTML coding, links were difficult to make. Since it took so much effort to link to something, one could reasonably argue that links were a reasonable proxy for authority—no one would go out of his way to link to crap, right?

Well, yes and no. Blogs took off in the late 1990s, making linking easy and unleashing the Force of the Many on the linked Web. While some argue that all linking has attenuated the value of a link, and therefore diluted the value of PageRank and other link-based relevance schemes, I believe that just the opposite is happening. Blogs are providing two crucial building blocks for the creation of a more intelligent Web.

First, blogs are personal statements by individuals, digital declarations of who they are and who they wish to be in the searchable world. Together with the ecosystem of links, both inbound and outbound, which grow around the specific site, the blog becomes a very nuanced (and eminently indexable) statement of individuals' social standing, relationships, interests, and history.

Second, once blogs reach critical mass (and I'd wager that has already occurred; we just don't know it yet), intelligent engines will be able to discern patterns among them that will provide second- and third-order relevance inputs that will help refine and return far better search results. Just as with folksonomies, it's Yahoo's early problem of trying to edit the Web solved by the Force of the Many. Human-edited classification schemes are far better than machines at discerning relevance, but they fail to scale to the size of the Web. But what if you used blogs as a proxy for thousands upon thousands of professional taxonomists?

A Glimpse of the Semantic Future

To garner a glimpse of the semantic Web in action, I drove down to IBM's Almaden research lab in San Jose, California. (To say the folks there are interested in blogs is an understatement.) The Almaden lab lies in a rather surreal juxtaposition with its surroundings. The center is sculpted into what must be at least a thousand acres of pristine Bay-area hillside; to get there, you must navigate three miles of uninhabited parkland. From the looks of it, it may as well be Norman Juster's *Phantom Tollbooth* (fittingly, at that). Nearby, Hollywood set-piece cows chew Hollywood set-piece cud.

The gate opens and you drive one-quarter mile to a four-story slate gray building, which looks rather like a Nakamichi preamp, only with windows (and landscaping). Inside are six hundred or so pure and applied researchers who are . . . well, mostly thinking about really difficult computer science problems. And this center is just one of eight that IBM supports around the globe. The others are in places like Haifa, Switzerland, Japan, China, and India. It's quite impressive, and reminds you that while the media can get carried away with one company at one moment in time, some firms have been hiring PhDs and putting their brains to good use for longer than most of us have been around.

I met with a couple of these scary smart guys, Daniel Gruhl and Andrew Tomkins, the lead architect and chief scientist, respectively, of IBM's WebFountain project. I'd heard a lot about WebFountain, and what I gathered sounded promising—it's been called an "analytics engine" by none other than the Institute of Electrical and Electronics Engineers (IEEE), the high holy council of geekery.

First, a bit of history. WebFountain is the offspring of nearly ten years of work at Almaden on the problem of search. That work began with Jonathan Kleinberg, the man who met with Larry Page back in the early days to swap notes on BackRub. Kleinberg agrees with the consensus view that search is in its early days. The really hard problems—natural language queries, for example—have yet to be solved. Search has gotten pretty sophisticated using keyword matching and link-pattern analysis, he notes. But search technology still has no idea what a document actually means—in the human sense.

WebFountain seeks to address this problem, and attacks it from two sides: first, by tagging the document itself with a top-down approach (more on that later), and second, by building what might be called the perfect query. A core problem with search as we know it is that of the inverse search. In an inverse search scenario, you intuit that there is a perfect query that, if typed into a search engine, would yield exactly the set of pages you're looking for. But you don't know the term, and your attempts to divine it continually bring up frustrating and irrelevant results.

Say, for example, you want to know more about that regulation you've heard about, the one that says you have the right to fly—with no additional charge—on a different airline if the one you are booked on cancels your flight. You want to find out the specifics of that regulation, but how?

You might Google "regulation airline overbooked" or something like that. That takes you to a few pages that are relevant—if you're in Europe. So maybe try it again, this time with a "Europe" (this tells Google to ignore pages with the word "Europe" in it—never mind that we're already way over the heads of most normal searchers' knowledge). Nope—at least not in the first few pages of results. Maybe take out all the EC and EU references? No again, but you have managed to waste five minutes reading a document by an obscure policy think tank that seemed promising, but didn't pan out. Frustrated, you probably give up—maybe it's time to call a research librarian or that friend of yours who worked at Delta.

But if you knew that the regulation was, in fact, called the FAA Rule 240, you'd be in like Flynn. That query gives you exactly the information you need. How might a computer learn to act more like a reference librarian and make the leap from "that regulation that lets me fly on other airlines" to "FAA Rule 240"? WebFountain is working on solving exactly that problem.

So Why WebFountain? Why Now?

IBM noticed that large companies were drowning in information and that broad search engines like Google were not providing relief. To deal with the complex information typically found in a large enterprise, corporate IT departments were trying to invent a new kind of mousetrap—one that solved a very specific, rules-based problem inherent to large corporations. But to invent this particular mousetrap, you needed more talent, resources, and hardware than any one organization could justify. Enter IBM.[7]

WebFountain is a classic IBM solution to the search problem. Instead of focusing on the consumer market and serving hundreds of millions of users and searches a day, WebFountain is a platform—middleware, in essence—around which large corporate clients connect, query, and develop applications. It serves a tiny fraction of the queries Google does, but my, the queries it serves can be mighty interesting.

Using WebFountain, for example, an IBM customer can posit a "theoretical" query such as this: "Give me all the documents on the Web that have at least one page of content in Arabic, are located in the Midwest, and are connected to at least two similar documents but are not connected to the official Al Jazeera Web site, and mention anyone on a specified list of suspected terrorists." Not the kind of query you'd punch into Google. (As to what kind of customer might want to be asking this kind of query, IBM is understandably mum. But it does stress that, hypothetically, these kinds of queries could certainly be asked of WebFountain by clients unstated.)

Another type of client might want to answer this kind of question: "Tell me all the places on the Web where *The Passion of the Christ* is discussed that also mention one of the top five box office movies that is not *Lord of the Rings,* and throw out all sites that either are in Spanish, or are in the Southern Hemisphere. Oh, and translate the ones that are not in English when you return results."

Could a global oil company find out what college students in the Bay Area are saying about the price of gasoline? Yup. Teenagers and fashion, mall-related zip codes? Done. Music label and artist buzz, so as to allocate a marketing budget? No problem (in fact, the idea for WebFountain sprang from just such a request).

So how does WebFountain make answers to such complex and specific queries possible? Short answer: a lot of hardware and a boatload of metadata tagging. Longer answer: WebFountain does more than index the Web, then serve up results based on keyword matches and some clever algorithms. Sure, it indexes the Web, but once the pages are crawled, WebFountain goes several steps beyond consumer search engines, classifying those pages across any number of semantic categories. WebFountain basically restructures the Web, making it accessible to a client's queries.

Just for fun, here's a partial list of how each and every Web page (or document, in IBM's terms) is annotated:

- Language
- Character encoding
- Porn (WebFountain has found that 30 percent of the Web is porn.)
- Duplicate status (Is it a duplicate or near duplicate of another page?)
- Date crawled
- Date of content
- Set of tokens (words) on the page
- Author (for selected document types)
- Source category (media site, major newspaper, etc.)
- List of entities on the page, where this can be a hierarchical set:
    - People
        - Government
        - Education
        - Business
    - Places (geolocation, including longitude and latitude)
    - Companies
    - Organizations

WebFountain can also tag entities on a page, creating sentiment around an entity, themes and associations for entities, and relationships between entities. Even more extraordinary, WebFountain customers can create entirely new tagging schemes, and IBM can crank the entire database—that'd be the entire Web—through those custom filters on the fly.

The Platform Play

As I mentioned earlier, IBM's model for WebFountain is platform-based. Almost anyone can develop for it (if he can pay the freight) using a standard programming interface that leverages simple Web services. IBM won't disclose most of its customers, but two it will mention are Semagix, which has a (pretty damn frightening) money-laundering detection application, and Factiva, which developed but later abandoned a "reputation manager"—a first-generation version of blog-based search.

With WebFountain, IBM has sliced the Web into subjective, structured data sets. It's created a search platform that allows a client to posit nuanced and entirely specific questions the answers to which may mean millions to that client, but are meaningless to most causal Web searchers. Hence, WebFountain will never scale to the reach of an application like Google.

Or, I wondered after I left IBM's facility, will it? I later asked Gruhl if there wasn't a point at which the power of WebFountain might be available to the greater Web community. Why not? After all, Overture and Google made it to billions in revenue 25 cents at a time; why not license WebFountain to an entrepreneurial company looking to beat Google at its own game, perhaps by placing a friendly interface on top of the WebFountain platform, and letting smaller companies and individuals get in on the party?

Gruhl thought about it for all of a millisecond, then said Moore's Law had not caught up to the computing demands of WebFountain, for now at least. All that annotation takes a lot of cycles and a lot of software, and the whole process must happen in a particular order. You can't throw more Linux boxes at the problem the way that Google does. Imagine if Google had to reindex the whole Web for each new searcher who uses the service. But Gruhl did admit that at some point in the future, WebFountain-like features may well scale to millions of queries a day. It's just a matter of time.

For now, WebFountain is your classic supercomputer application, though in this case, the supercomputer consists of 256 dual-processor blades attached to well north of half a petabyte of storage. Compared with Google, it has far fewer processors banging away, but the throughput is "in the top fifty of all supercomputers on earth," Gruhl says quite proudly. In other words, the entire Web can be scarfed up, tagged, and retagged in less than twenty-four hours. Because of the distributed nature of its computing architecture, the process of updating Google's entire index takes nearly a month (though portions are now updated far more frequently).

But it seems to me the two companies, as distinct as they are, are racing toward a middle where they may well meet. Google and most other consumer-facing search engines are obsessively focused on understanding user intent—on deriving the most relevant results, regardless of how vague a query might be. This is because folks usually come to Google with poorly structured intentions—most searchers ignore the advanced search features and use just two or three words per query. Further, Google's indexing process relies on scalable but unstructured approaches to keyword matching and link analysis. Despite these limitations, the pressure to innovate is intense, and the PhDs at the Googleplex will continue to innovate, cooking up new hacks to bring the Web to heel.

The folks at IBM, on the other end, having brought the Web (somewhat) to heel, have created a platform that developers can increasingly exploit in larger and more profitable markets. But the query language is complex and unapproachable to consumers, and the back end is cumbersome to say the least. Might we someday get a GoogleFountain? I certainly hope so, and suspect it's only a matter of the future catching up to our present. The computer on which I'm writing this book is the direct descendant of a 1960s-vintage supercomputer that was once locked away in a supercooled nerve center, just as WebFountain is now. Can you imagine the day when anyone with a Web connection can query WebFountain, in a format as ubiquitous, intuitive, and well mannered as Google? That's a pretty strong step toward perfect search.

Federated and Domain Specific: Focus, Focus, Focus

But if we're going to get to perfect search, we might think about taking baby steps first. Enter domain-specific search. Domain-specific search solutions focus on one area of knowledge, creating customized search experiences that, because of the domain's limited corpus and clear relationships between concepts, provide extremely relevant results for searchers.

A good example of this is GlobalSpec, an engineering-specific search engine that got its start in the mid-1990s as an online catalog. The site basically moved all catalog-based information about engineering parts—sensors, transducers, accelerometers, and so on—into a huge, cross-referenced database, which it then distributed over the Web. The idea was not exactly innovative: make money by connecting customers to parts suppliers over the Internet. Simple.

Over the years GlobalSpec evolved into a robust community of a million or so engineering types who use it to find and spec parts. That alone is pretty cool (I mean, a million engineers!). But in early 2004, GlobalSpec realized that while it had a good catalog and a great user base, it didn't have the ability to easily answer all the questions its community might come up with, and it was losing potential customers to general search engines like Google.

Following the maxim that search drives commerce, GlobalSpec's executive team came up with a focused search product they call the Engineering Web. In essence, GlobalSpec's human editors identified 100,000 or so very specific sites that they believed contained information related to the domain of engineering. They then built a crawler that indexed just those sites (and related sites they found through their crawl, of course). Then GlobalSpec took its crawl one step further. Not only did it crawl the public engineering Web; it also surfaced invisible Web databases not found in mainstream search engines—patent and standards sites, for example, which are walled off by registration and business considerations.

Presto: a domain-specific search engine that, while not perfect, pretty much kicks Google's butt on one (admittedly narrow) subject.

Because of its limited domain, GlobalSpec can use relatively simple keyword-based algorithms to surface lists of ideas or terms related to your search. This allows you to refine your search in ways that simply don't scale in the Googleverse. These related ideas are inferred from the results of your initial query. For example, if you search on "aerodynamics," you will get related subject searches for "aircraft, flight mechanics, helicopter aerodynamics, computational fluid dynamics, and theoretical aerodynamics" as well.

This is clustering—a technique used by major search engines like Ask Jeeves, AOL, and others—but with far superior results. When you live in a gated community of domain specificity, you weed out the riffraff of false positives which roam the big bad Web.[8]

Because anyone can use the service—it's not limited to registered users—GlobalSpec has created a portal that drives traffic and intent through its original database business, and in the process it has built an intelligent island of engineering information that lives in the public sphere.

Sure, you probably don't usually spend a lot of time comparing accelerometer specifications, so why should you care? To my mind, GlobalSpec points the way toward the creation of untold numbers of powerful vertical search engines, engines which, because they are limited in domain and exclusive by nature, can, in fact, offer extremely cool tools to find exactly what you want.[9] The commercial payoff of search is driving more and more entrepreneurs to polish out significant portions of the Web with semantic-like functionality. And when the borders of those engines begin to touch each other, lily-pad like, magic can happen.

Circling back to our goal of perfect search, imagine that nearly every subject worthy of some critical mass of human intent—from archaeology to automobiles, zoology to pop music—receives a GlobalSpec-like vertical search treatment, or perhaps a blog ecology that serves as a useful proxy. Then imagine engines like Google and Yahoo crawling each of them, and creating something of a metasearch engine on top of hundreds or thousands of domain-specific sites. It's not such a leap to imagine, in the end, that we get closer to perfect search through the concerted efforts of thousands of smaller sites making their domain more perfect.

There are plenty of signs pointing this way already. Metasearch is a thriving industry, mainly because two of the three pieces of search—crawling and indexing—have already been done by someone one else. And domain-specific sites are slowly but steadily launching, with the most commercial of them coming off first.[10]

It's not hard to imagine that as domain-specific search sites proliferate, so will federated or metasearch sites, specializing in taking your relatively inchoate query and guiding you through layers of results to your perfect answer.

The Web Time Axis

A UC Berkeley study reported that in 2002, the most current year for which there are figures, humankind created 5 exabytes of stored data—in paper, the equivalent of creating 500,000 new Libraries of Congress each year. By stored data, the Berkeley scholars meant print, film, and optical media (hard drives). More than 90 percent of those 5 exabytes were stored on a hard disk—a device that didn't exist just sixty years ago. Every day we create and store more information (in digital format) than had been stored for most of our history on paper.

But as we know, for the most part all that information is not available to most search engines. The invisible Web is one major reason why, but another has to do with the nature of the Web itself: every time a Web page changes, or goes off-line, the original version is lost.

In short, the Web has no memory. Want to read thestandard.com from back in 1999, during the height of the bubble? I would, too, but you won't find it in Google's index. Want to find the first-ever list of Jerry and David's Guide to the World Wide Web? So would Yang, but he never kept a copy.

But at some point in the not too distant future we'll have live and continuous historical copies of the Web that will be searchable—creating, if you will, a time axis for the Web, a real-time Internet archive with a copy of the Web for every day of the year, and every year in perpetuity. In other words, in our lifetimes we'll see our cultural digital memory—as we understand it through the Web and engines like Google—become contiguous, available, always there. And barring a revival of the Luddites or total nuclear war, this chain will most likely be unbroken, forever, into the future.

Historians looking back to this era will mark it as a watershed. At some definable point in the twenty-first century, the Web will gain a memory of itself, one that likely will never be lost again. This will probably start as a feature of a massively scaled company like Yahoo, Google, or Amazon. But it's coming, and the implications are rather expansive.

If the Web had a time axis, you could search constrained by date. You could ask questions like "Show me all results for my query from this time period" or "Tell me what were the most popular results for 'George W. Bush' on May 3, 2004." How about "Show me every reference to my great-grandfather during 2006?" In the future, your great-grandson will probably do just that. Thanks to the dramatic decrease in the cost of storage, the dramatic increase in computing power, and the scalable business model of paid search, this day is not far off. The Web is just ten years old, for the most part, but think what it might be like when it's one hundred years old. That's a lot of data to search, and a lot of opportunity for innovation.

But can we realistically expect the ability to search by time? So far, the challenge is daunting. That's because while it's true bits can be eternal, so far we have not done much to ensure that the infrastructure of the Web takes advantage of that fact.

If search is going to be perfect, then we need to be able to access the world's knowledge. Brewster Kahle is trying to address this problem by creating a massive nonprofit project that is attempting to archive both print and film, as well as the entire Web, as best as it can, on a nearly daily basis. Called the Internet Archive, the project has been spidering and archiving the Web every day since 1996. As Brewster told me when we discussed archiving at his San Francisco offices, "The lesson of the first library of Alexandria is 'don't have just one copy.' "[11]

Kahle is something of a folk hero in search, having started WAIS, an early Internet publishing and search service, and Alexa, a still innovative search company purchased in the late 1990s by Amazon. Alexa was one of the first companies to use a consumer toolbar to track clickstream data, and remains a key part of Amazon's A9 search service.

But if we are ever going to realize the potential of the Web time axis, we'll need the Force of the Many out there making copies of the Web over time, and archiving them in such a way that we can all get access. (The Internet Archive can do only so much.) The glimmerings of such an ecosystem are appearing all around us. Personalized search history is one such development. So is LookSmart's Furl tool, which allows you to take the equivalent of a Xerox copy of any site you visit, then store it for future sharing, searching, and viewing. Ask Jeeves announced a similar service in late 2004, and Google, Yahoo, and AOL will most likely have comparable services available by the time you're reading this.

When a good portion of the general public gets in the habit of saving and sharing Web pages, and those pages are saved forever, someone will come up with the idea of folks "donating" copies of their pages to some kind of universal Web memory project. Examples of similar projects already abound on the Web: the *Wikipedia,* a volunteer-edited encyclopedia, surpassed 1 million articles in September 2004, and nearly all search engines use DMOZ, a volunteer-edited Web directory. Once such a project gets going, volunteers will probably start copying vast parts of the Web into such an archive, most likely in their spare time, out of self-interest (*I want to make sure my site is archived forever!*) as well as for the enlightened greater good. And once vast portions of the past Web have been archived, engines like Google and Yahoo will certainly index them, bringing the Web time axis online, for good.

The Search for Perfection

We've bitten off a lot in this chapter, so let me try to summarize. I posited the rather blue-sky notion of perfect search, and then broke down a number of trends that are pointing toward fulfilling at least some part of that larger vision. Those trends are ubiquity (the integration of more and more information into Web indexes), personalized search (the application of your personal Web toward a more perfect answer), the rise of the semantic Web (the tagging of information so as to make it more easily found), domain-specific search, and the Web time axis. But how does it all fit together?

Google aside, there's no single moment when all these trends converge. Think back to your first Google epiphany, or if you've been searching the Web for a while, your first AltaVista epiphany. Think about what that felt like—how all of a sudden you realized the world was, quite literally, at your feet (or rather, your fingertips). Perhaps it was the first time you entered your own name into Google and realized that the world saw you as the sum of those results. Or maybe it was the time you found the perfect CD because of a recommendation made by Amazon's search algorithms. Or maybe it was the first time you installed a desktop search program and found that obscure e-mail thread you'd forgotten about. Or maybe it was the first time you used Google's video search to find the next airing of your favorite show, and realized that very soon, you'd be receiving most of your television programming over the Web.

Whatever your first perfect search moment was, there will be many, many more as the space evolves. Search is no longer a standalone application, a useful but impersonal tool for finding something on a new medium called the World Wide Web. Increasingly, search is our mechanism for how we understand ourselves, our world, and our place within it. It's how we navigate the one infinite resource that drives human culture: knowledge. Perfect search—every single possible bit of information at our fingertips, perfectly contexualized, perfectly personalized—may never be realized. But the journey to find out if it just might be is certainly going to be fun.

Epilogue

Search and Immortality

On a fine sunny morning in 2003, not long after the birth of my third and most likely final child, I typed "immortality" into Google and hit the "I'm feeling lucky" button. I can't explain why I turned to a search engine for metaphysical comfort, but I sensed the search might lead me somewhere—here I was writing a book about search, but what did it matter, really, in the larger scheme of things?

In an instant, Google took me to the Immortality Institute, an organization dedicated to "conquering the blight of involuntary death."

Not quite what I was looking for. So I hit the search again, but this time I took a look at the first ten results, etched in blue, green, and black against Google's eternal white. Nothing really caught my eye. Cryonics stuff, a business called Immortality Inc., pretty much what you might expect. I couldn't put what I was looking for into words, but I knew this wasn't it.

Then I noticed the advertising relegated to the right side of the screen.

There were four ads, each no more than three lines of text. The first was someone who claimed to have met immortal ETs. Pass. The third and fourth were from eBay and Yahoo Shopping. These megasites had purchased the immortality keyword in some odd and obliquely interesting hope that people searching for immortality might well find relief through . . . buying shit online. (In fact, what Yahoo and eBay were doing was the equivalent of search arbitrage—buying top positions for a search term on Google and then creating a link to the exact same search term on their own sites, in the hope of capturing high-value customers). Interesting, but I wasn't looking to *buy* the concept of immortality; I wanted to *understand* it. I took a pass on those as well.

But the second paid link pointed to the epic *Gilgamesh,* which I hazily recalled as the first story ever written down—in Sumerian cuneiform, if memory served. I clicked on the link, earning Google a few pennies in the process, and landed on an obscure bookseller's page. The epic of *Gilgamesh,* the site instructed me, recounts mankind's "longing stretch toward the infinite" and its "reluctant embrace of the temporal. This is the eternal lot of mankind."

Bingo. I didn't quite know why, but this was the stuff I was looking for. My vague desire to understand the concept of immortality had brought me to the epic of *Gilgamesh,* and now I was hooked. My search was bearing fruit.

But I didn't want to buy a book and wait for it to come. I was in the moment of discovery, the heat of possible consummation. I wanted to read that epic, *right now.*[1] So I typed the title itself into Google, and once again found myself larded with options. But this time the organic results (the search results in the middle of a Google page, as opposed to the ads on the right) nailed it: the first two offered direct translations of the stone tablets upon which the epic is written. Clicking on the first link, I found a Washington State University professor's summary of the Gilgamesh story.

Gilgamesh, I learned, was the king of a place called Uruk in ancient Babylonia (in what is now Iraq). The professor, Richard Hooker, explained that the civilizations in that area, among the first known to man, developed many legends around the king, as much to explain their own society as the man himself. The first of these was recorded around 2500 B.C. on clay tablets in cuneiform. Hooker continued:

The fullest surviving version . . . is derived from twelve stone tablets . . . found in the ruins of the library of Ashurbanipal, king of Assyria, 669–633 B.C., *at Nineveh. The library was destroyed by the Persians in 612* B.C. . . . *The tablets actually name an author, which is extremely rare in the ancient world, for this particular version of the story: Shin-eqi-unninni. You are being introduced here to the oldest known human author we can name by name!*[2]

In my search for immortality, I had found the oldest known named author in the history of Western civilization. Thanks to the speed, vastness, and evanescent power of Google, I came to know his name and his work within thirty seconds of proffering a vaguely worded query. This man, Shin-eqi-unninni, now lived in my own mind. Through his writings, with an assist from Google and a university professor, he had, in a sense, become immortal.

But wait! There's more. Gilgamesh's story is one of man's struggle with the concept of immortality, and the story itself was nearly lost in an act of literary vandalism—the destruction of a great king's library.

As I contemplated all of this, sensing that, just possibly, I had found a way to explain why search was so important to our culture, I read the first tablet's opening lines:

The one who saw all (Sha nagba imuru) I will declare
to the world. . . .
He saw the great Mystery, he knew the Hidden:
He recovered the knowledge of all the times before the Flood.
He journeyed beyond the distant, he journeyed beyond
exhaustion,
And then carved his story on stone.

What does it mean, I wondered, to become immortal through words pressed in clay—or, as was the case here, through words formed in bits and transferred over the Web? Is that not what every person longs for—what Odysseus chose over Kalypso's nameless immortality—to die, but to be known forever? And does not search offer the same immortal imprint: is not existing forever in the indexes of Google and others the modern-day equivalent of carving our stories into stone? For anyone who has ever written his own name into a search box and anxiously awaited the results, I believe the answer is yes.

Acknowledgments

Writing is a solitary act, but I've found that authoring a book requires a community. This book would be far poorer (and most likely still unfinished) were it not for my family, my friends, and my colleagues.

Not many editors would see the value in a book on search, given the ugly state of the Internet industry in early 2003, but Adrian Zackheim did, and for that I am deeply in his debt. And not many agents would do the same, but Esther Newberg at ICM not only saw the book when I pitched it to her; she took me on as a client, even though I had never written so much as a foreword.

Against the better advice of most of my friends, I wrote *The Search* at home. I took great comfort in the daily routine of my three children and my extraordinary wife, Michelle, to whom this book is dedicated. Not only did Michelle manage to keep our home together while I cursed and fumbled in my study; she made it thrive, creating a backdrop that informed and enlivened my work. She also read the early drafts and offered the first words of encouragement (and gentle criticism).

As I began the process of reporting, I thought that starting a weblog chronicling my research might be interesting to a few people, and if I was lucky, a source or two might take pity on my ignorance and raise a hand to help. I had no idea that the site—Searchblog—would become not only my daily obsession but an essential part of my writing and reporting life. If any praise is attached to this book, it will have its root at Searchblog with the tens of thousands of readers who challenged me, corrected me, and encouraged me to continue. Searchblog and its readers offered me a place to try out new ideas, ask for help on sticky reporting problems, and test early drafts of the book. I have no idea how authors manage to cope without such support. And I must also thank Scot Hacker, Searchblog's system admin, who has managed the site's growth (and my technical handicaps) with grace and humor.

As I struggled to understand what it meant to write a narrative, many kind souls offered advice and aid. At the Graduate School of Journalism of the University of California, Berkeley, Dean Orville Schell offered not only a place to hang my hat but a network of intelligent and accomplished faculty and students who provided critical guidance in the early stages of the book. The inestimable Clay Felker, my partner in teaching and my mentor in publishing, never failed to offer assistance. I am also indebted to my first two research assistants, Ben Temchine and Mary Nicole Nazzaro, both of whom have gone on to greater glory in the real world since graduating from the J School. The extremely patient and able Abigail Phillips helped me in the final stages of my research, as did Stefanie Olsen, who provided her insight into the history of search.

I am grateful to the industry experts who tutored me and kept me honest—in particular Gary Price of Resourceshelf and Search Engine Watch, and Danny Sullivan, also of Search Engine Watch. Both read portions of the manuscript and gave me invaluable advice. Among the hundreds of industry insiders who suffered my often naive questions, these stand out: Chris Anderson, Stewart Baker, Andy Beal, Gordon Bell, Jeff Bezos, Tim Bray, Brett Bullington, Stewart Butterfield, Dick Costolo, Barry Diller, Mark Fletcher, Danny Hillis, Mike Homer, Bill Joy, Vinod Khosla, Matt Koll, Joe Kraus, Steve Krause, Kevin Lee, Philipp Lenssen, Greg Linden, Udi Manber, Mary Meeker, Halsey

Minor, Neil Moncrief, Louis Monier, Scott Moore, Mike Moritz, Martin Nisenholtz, Joyce Park, Scott Rafer, Safa Rashtchy, David Sifry, Graham Spencer, Raymie Stata, Jonathan Weber, Jake Winebaum, and far too many others to mention.

As this narrative turns on several key actors, I owe special thanks to the people who helped me coordinate scores of interviews at the principal companies, Yahoo (including Overture and AltaVista) and Google. At Yahoo, my thanks go to Chris Castro, and at Google, to Cindy McCaffrey and David Krane. All of them are far too busy to put up with the incessant requests I made over the past two years, but somehow they found time for me, over and over again. Larry Page and Sergey Brin at Google, and Jerry Yang and David Filo at Yahoo, were extremely kind with their time, as was Bill Gross at IdeaLab. Along those lines, without the advice of Ted Meisel, Dan Rosensweig, Jeff Wiener, Jeremy Zawodny, and many others at Yahoo, or of Patrick Keane, Steve Langdon, Marissa Mayer, Megan Smith, Susan Wojcicki, and scores of others at Google, I would have been lost. Similarly, I owe Steve Berkowitz and Jim Lanzone at Ask a debt of gratitude, as well as Yusuf Medhi, Gary Flake, and David Cole at Microsoft.

As I began writing in earnest, several stalwarts came to my rescue. My mother, Priscilla Battelle, shared her knowledge of literature in general and Greek mythology in particular. My father, Richard Battelle, and sister, Ann Bool, always encouraged me, regardless of my sporadic and paltry attempts at communication during my self-imposed hibernation. Denise Caruso and I shared the unique pain of authors who struggle to make, but ultimately miss, their deadlines. Douglas Rosenberg offered early reads and suggestions, and Josh Quittner, my editor at Business 2.0, managed to both support me and look the other way as my allegiances to his magazine and my first book were tested. In a similar vein, I owe Tim O'Reilly, Dale Dougherty, Mark Jacobsen, Gina Blaber, and the O'Reilly and Media Live teams deep thanks for giving me the chance to launch the Web 2.0 conference even as I wrote the book.

The same goes for the band at Boing Boing—many thanks to Mark Frauenfelder, David Pescovitz, Xeni Jardin, and Cory Doctorow.

As the smoke cleared and a manuscript emerged, I turned to Bill Brazell for initial line editing. Should he ever decide to offer his services to the public, he'll never want for work, for he is without peer. And then there's John Heilemann. A friend for over a decade and partner in many a venture, John spent countless hours on the phone with me, demanding that I do better, forcing me to acknowledge every error in structure, each lapse in rigor, every lazy cliché. I left a few in just to piss him off, but I cannot imagine the book without his exacting friendship.

Once I had the courage to turn the book over to Adrian Zackheim, his edits were lucid and deft, and the members of his team—in particular Megan Casey, Will Weisser, and Allison Sweet—were not only professional; they were fun to work with, shattering for me the myth that publishers were a stodgy and querulous lot.

I know I have left out countless others, so please accept my apologies in advance. Nearly four hundred people were generous enough to sit for interviews during the course of my research, and only a small percentage of them appear by name in the final work. But if the book is spoken of well by anyone, it is because of their generosity.

Finally, I must acknowledge the reader of the book itself, because I view this as a living work, one shaped by the reader as much as the author. I am quite sure there will be errors and omissions in this volume, and the pace of change in the search space guarantees that this book will be somewhat out of date by the time it is read. I am committed to updating this work at the Searchblog site. Those readers who care to are invited to head over to www.battellemedia.com/thesearch, where it is my hope we can continue the conversation apace.

John Battelle
Kentfield, California
May 2005

Notes

1. The Database of Intentions

1. In the summer of 2001, my business, the high-flying Standard Media, parent of *The Industry Standard,* had imploded in spectacular fashion. Like hundreds of other Internet companies, the Standard fell victim to uneven management decisions, a terrible market, and eviscerating battles at the board level over what to do about it. The experience left my optimism and enthusiasm for the Internet sorely wanting.
2. The first-ever iteration of the Google Zeitgeist is at http://www.google.com/press/zeitgeist.html.
3. From 1992 to 1997, I was a cofounding editor of *Wired;* from 1997 to 2001, I founded and ran Standard Media. In the fall of 1998, *The Industry Standard* was the first business magazine to put Google on the cover (along with three other search engines under the title "Search That Works"). I knew about Google, certainly, but it took another three years for me to realize its true importance.
4. John Poindexter, famous for his role in the Iran-Contra scandals of the 1980s, resurfaced as a special adviser to the Pentagon in 2002. His dream, according to published reports, was to create a huge government database that would monitor every possible source of information, including the Internet, so as to alert the government to possible terrorist threats. Its very name—Total Information Awareness—sparked a backlash, and the program lost its public funding in 2003. However, portions of the program live on in various defense and intelligence agency budgets.
5. Social networking—which you might call a people search application—has received a significant amount of venture capital investment and software

development in recent years. By mid-2003 one in ten Internet users had registered at a social network, and one in five had visited such a network.

6. There's more on Google PageRank in Chapter 4.
7. For more on the USA PATRIOT Act, head to Chapter 8.
8. As this information has become eternal, we, as creators of that information, have lost a large degree of control over how it is used and in what context. In fact, in many cases we have lost ownership of the information altogether—arguably before we even knew it existed in the first place. Whether this matters at all is worth debate: after all, how could we lose that which we never had? It's not my goal to write a privacy screed, or take "evil corporations" to task. But it seems to me the issues raised by the ownership of our collective data exhaust are certainly worth raising and discussing, with a particular eye toward the law of unintended consequences, if nothing else. As we move our data from the edges to the center, a question arises: have our assumptions moved with our data?
9. For a good example, head over to www.alicebot.org/.

2. Who, What, Where, Why, When, and How (Much)

1. Many thanks and sincere gratitude go to Gary Price and Danny Sullivan of Search Engine Watch for some of the examples used in this section.
2. A good place to start is Tara Calishain's excellent site ResearchBuzz.
3. Majestic Research can be found at majesticresearch.com.
4. One should note that the Kelsey Group has something of a stake in claiming local search numbers as high as this. It's a research group that focuses on the yellow pages market.
5. Brin and Page, "The Anatomy of a Large-Scale Hypertextual Web Search Engine."
6. Turns out just this approach will be tried by television. For more on that, turn to Chapters 7 and 10.

3. Search Before Google

1. Early in eBay's history, founder Pierre Omidyar was often quoted as saying that he started the company so as to help his wife sell Pez dispensers. The truth is far less anecdotal—Omidyar simply wanted to make the Internet more useful and efficient.
2. Reid later went on to work at Google, and he has more than passing experience with large companies and controversy, as his pending lawsuit against Google illustrates. For more on that, head to Chapter 9.

3. For a great history of this story, pick up Douglas Smith and Robert Alexander's *Fumbling the Future: How Xerox Invented, Then Ignored, the First Personal Computer* (Lincoln, NE: toExcel, 1999).
4. As an early cofounder of Wired Digital, I profited from this sale.
5. Rob Reid, *Architects of the Web* (New York: Wiley, 1997), page 241.
6. I was a senior manager at Wired Ventures, parent of HotWired, and I must confess, at the time I thought starting a search engine was a rather loony idea.
7. Many have noted that Google built its business on the back of Yahoo, much as Yahoo built its business on Netscape. What is not well known is that when Tim Koogle left Yahoo in the spring of 2001, he encouraged new CEO Terry Semel to meet with Google's founders, Larry Page and Sergey Brin. Koogle felt Yahoo should own search, and buying the wildly popular but revenue-deficient Google seemed a perfect way to do it. But "there was no chemistry between Terry and Larry and Sergey," said an executive close to both companies. Yahoo ultimately did end up buying its way into the search game (it purchased Inktomi, AltaVista, and Overture), and it now stands as Google's most significant competitor in the space.

4. Google Is Born

1. In the early 1990s, Gates cast a tremendous shadow across his then young company. In *Microserfs,* Douglas Coupland's deeply reported but fictionalized account of life at Microsoft in the early 1990s, engineers told tales of purposely walking in front of Gates's office window in an attempt to show the founder that they could get from one place to another with the least number of steps. Gates has been known for being dismissive of Google in private—but he certainly can't help but admire what the company has accomplished.
2. From the online encyclopedia *Wikipedia:* "In mathematics and computer science, a graph is a generalization of the simple concept of a set of dots, called vertices or nodes, connected by links, edges or arcs. Depending on the applications, edges may or may not have a direction; edges joining a vertex to itself may or may not be allowed, and vertices and/or edges may be assigned labels. A numeric label is often called a weight. If the edges have a direction associated with them (indicated by an arrow in the graphical representation) we have a directed graph. This means it is possible to follow a path from one vertex to another, but not in the opposite direction. If there are no directed edges, the graph is an undirected graph. Unless otherwise indicated, the term graph typically is assumed to mean a simple

graph, in which at most one edge exists between any two vertices (directed or undirected)."

3. As noted later in the chapter, Kleinberg's work on hubs and authorities is considered seminal. Page and Brin's original paper outlining Google, "The Anatomy of a Large-Scale Hypertextual Web Search Engine," cites Kleinberg's work. In a practice common to all academic writing (and, not coincidentally, most bloggers), Kleinberg returns the favor, citing Brin and Page's work in his own works. Klienberg's work also informs IBM's Web Fountain (see Chapter 11) and Ask Jeeves's Teoma technology, among others. Page and Brin's original paper can be found at http://www.db.stanford.edu/~backrub/google.html.
4. From an August 1991 Usenet post by Tim Berners-Lee announcing the WWW: "The project started with the philosophy that much academic information should be freely available to anyone. . . . The WWW world consists of documents, and links. Indexes are special documents which, rather than being read, may be searched. The result of such a search is another ('virtual') document containing links to the documents found. A simple protocol ('HTTP') is used to allow a browser program to request a keyword search by a remote information server."
5. One of the first to do it was Louis Monier, who launched AltaVista in 1995 using the resources of Digital Equipment Corporation (see Chapter 3).
6. Early projects which caught Brin's interest included determining a method by which previously shredded paper documents could be reconstituted, as well as designing a system that would give copyright owners the ability to distribute their property in digital format. That problem has yet to be solved to the satisfaction of most in the content business.
7. When I asked Steremberg about the inspiration for BackRub, his first response was "I think Larry just wanted to find out who was linking to him."
8. In their paper on PageRank, Page and Brin show an example of such a search for the word "university" and compare results with that of the top engine at the time—AltaVista. The difference in quality is irrefutable.
9. It's worth noting that as of early 2005, the top-ranked site for "Ulysses S. Grant" is now the very site that elicited the anguished e-mail back in 1998. Clearly, the Webmaster figured out how to get into the good graces of Google's index.
10. It's certainly the second most cited paper, at least by the count of information retrieval expert Lee Giles at Penn State. His Citeseer service counts 457 citations to "Authoritative Sources," ranking it just below Brin and Page's paper introducing Google, which had 499 citations as of December 2004.

11. Around this time, Page and Brin sought out the missing Beatles, Alan Steremberg and Scott Hassan, and granted them equity in the newly formed company. They did the same for Stanford University.
12. After the VC round, Bloomberg News called to interview Brin. Brin offered Bloomberg what would become one of his last public statements about going public. From the Bloomberg piece: "The investments from Kleiner Perkins and Sequoia mark Google.com's first substantial round of venture-capital funding. The company wants to sell shares to the public before raising more capital, said President Sergey Brin. He declined to give an expected date for an initial public offering."

5. A Billion Dollars, One Nickel at a Time: The Internet Gets a New Business Model

1. Magellan presaged a current boom in what is called desktop search. By the end of 2004, Yahoo, Microsoft, Google, AOL, Ask, and many others introduced search applications that scan a user's hard drive and make those results available in a Web-like interface.
2. In fact, traffic metrics—the number of pageviews or visitors a site garnered—became accepted shorthand for the process of valuation of Internet companies. A major problem with this approach was there was no well understood way of determining whether the traffic, in fact, would convert to paying customers of one kind or another.
3. Similar pricing models are now being floated to solve the e-mail spam problem.
4. I was there to launch *The Industry Standard,* but that's another story. We shared a stage, in fact, and most folks thought *The Standard* was a far more viable idea. Shows what they know.
5. Echoes of this disdainful refrain still resonate today: Google continues to make hay on Yahoo's practice of paid inclusion—mixing paid results into otherwise pure search listings.

6. Google 2000–2004: Zero to $3 Billion in Five Years

1. How I found this particular quote is in itself a story of Google's reach and power. As I was writing a passage about the famous "Don't Be Evil" motto and pondering the consequence of painting oneself into such a philosophical and moral corner, Orbital's "You Lot" rotated into first position on my iTunes. The tune turns on a sample of a stern lecture given by a British man about the power of technology (it's a bit reminiscent of Pink Floyd's *The Wall*). In any case, the quote struck me as aptly summing up the

quandary in which Google finds itself, so I punched "You Lot" and "Orbital" into my search box. Not much, so I added the album name, and I found a review. Reading through it, I found where the lecturing voice came from: it belongs to Christopher Eccleston, a well-known British actor, playing the Son of God in a rather obscure but well-reviewed television series called *The Second Coming.*

2. At an *Industry Standard* conference in the summer of 1999, investor and board member John Doerr was asked by journalist John Heilemann how Google was going to make money. Doerr cited Google's impressive traffic numbers—4 million pageviews per day, at that point—and said, quite accurately, "We'll figure out how to monetize that." That comment became something of a shorthand for legions of dot-com entrepreneurs chasing investment dollars during the bubble era.
3. GoTo later sued Google for patent infringement. The lawsuit was settled years later, when GoTo, now known as the Overture unit of Yahoo, dropped charges after Google proffered a multi-hundred-million-dollar payment.
4. GoogleScout later became the "related page" feature on Google.
5. At one point, Wojcicki told me, Google hired a marketing consultant who held focus groups to determine if Google should become a portal. Page came to the sessions and watched from behind a two-way mirror as the focus group participants discussed a portalized version of Google. He came away more convinced than ever that Google should remain pure.
6. Page and Brin both acknowledge that Google's approach to management has caused strains for some employees, and in 2004, with Eric Schmidt's help, they began to add additional layers of management.
7. Richard Wiggins, "The Effects of September 11 on the Leading Search Engine," in *First Monday,* an online journal. www.firstmonday.org/issues/issue6_10/wiggins/.
8. Why did the press love Google so much? My own theory stems from the fact that members of the press were early adopters of Google's service. Journalists are by nature eager to dig through reams of information to find that one fact, that one smart point of view. Put simply, Google's technology made it much easier for journalists to do their job.
9. Students of technology company history are probably experiencing a bit of déjà vu at this point—the scenes described herein are identical to those at Microsoft in the early 1990s, or Apple in the late 1980s. Certainly Google did not own the patent on high-tech geek heaven.

10. At the time, Mayer was reportedly Larry Page's girlfriend, a fact that conferred upon her actions even more cultural significance inside Google.

7. The Search Economy

1. By far the most compelling article for further reading on the long tail phenomenon is Chris Anderson's *Wired* article by the same name. Go to www.wired.com/wired/archive/12.10/tail.html.
2. For more on Google's Webmaster guidelines, go to www.google.com/webmasters/guidelines.html.
3. The full story on the eBay affiliate fraud is at www.auctionbytes.com/cab/abn/y03/m10/i24/s01.
4. JupiterResearch expects digital music sales to grow to $1.7 billion by 2009—12 percent of the total consumer music spending.
5. Because of the effect of search and blogging, I find that I read more things that have been pointed to by others, rather than only those that I pull down myself or that are pushed at me by publishers.
6. In a sign that times are changing, the *Journal* experimented with opening its site up to all comers for one week in late 2004. It has since opened up a sample of its daily stories to deep linking and hired an editor to oversee the paper's relationship with the blogosphere.
7. A far cry from early Internet commerce. The first version of e-commerce pioneer Amazon launched without a search box. "Shows what we knew," Jeff Bezos told me.
8. Yes, the same Warren Buffett who inspired the Google founders' "Owner's Guide" letter in the S1. For more, see Chapter 9.
9. But Google, as of mid-2005, still had a long way to go in this department, according to several midsize advertisers I spoke with. According to these advertisers, who spent from $50,000 to $150,000 with the company, Google rarely answered the phone and responded slowly, if at all, to their complaints of rampant click fraud.

8. Search, Privacy, Government, and Evil

1. This may change if search indeed becomes the place where true artificial intelligence arises, as outlined in Chapter 1.
2. The act's full name is Uniting and Strengthening America by Providing Appropriate Tools Required to Intercept and Obstruct Terrorism Act of 2001. I owe a debt to the Electronic Privacy Information Center (EPIC) for its analysis of the act and its implications. Its URL is www.epic.org/privacy/terrorism/usapatriot/.

3. And the list of institutions the government can query for your private information is growing. On the same day day that U.S. forces captured Saddam Hussein—a Saturday—President Bush signed the Intelligence Authorization Act for Fiscal Year 2004 into law. Given that the news cycle was focused on pictures of Hussein's oral health, most missed the fact that the act redefined the kind of information government authorities may intercept to include "financial information" from any business that might "have a high degree of usefulness" to FBI investigations. Combine this with PATRIOT, and pretty much everything you buy can now be reviewed by the government. The law was recently challenged in court and is under review.
4. Under the Foreign Intelligence Surveillance Act of 1978 (FISA), the government could spy only on "foreign powers" or "foreign agents." If the person the government wished to spy on was a U.S. citizen, the act required probable cause and belief that the person was engaged in espionage.
5. For more on the issues of corporate and government privacy, pick up a copy of Robert O'Harrow Jr.'s *No Place to Hide,* from Free Press.
6. Google is hardly alone—most corporations have privacy policies that give them great latitude.
7. Readers interested in understanding Google's China dilemma in a historical context would do well to read Schoenberger's book. It is a fascinating tale of an idealistic company—Levi Strauss—founded during the first California gold rush, whose noble principles ran headlong into the reality of China and the global outsourcing economy.
8. Schell referenced an appropriate Chinese expression: "To want to be a prostitute and erect a memorial arch to feminine virtue at the same time." (*Youyao dang biaozi, youxiang li paifang.*)

9. Google Goes Public

1. In an odd twist of fate, Google did lock up many of its employees from selling—though for an unusually short period of time. Many employees were livid that they could not sell at the offering, but they soon got over it when the stock skyrocketed and the lockups expired.
2. That Google would claim the status of an editorially driven company is interesting, given its reluctance to act like an editorially driven company in the context of its approach to organic search results. For more on this, read down to the section "The Competition."
3. To address some of its perceived governance shortcomings, the company did add three directors to its board: John Hennessy, president of Stan-

ford; Paul Otellini, president of Intel; and Arthur Levinson, CEO of Genentech.

4. Google's lockdown mentality hardened further once the CEO of salesforce.com, another high-profile IPO candidate, gave an interview to the *New York Times* during his company's quiet period, drawing an SEC rebuke and subsequent delay for its IPO.
5. In Homer's Greece, the Fates were represented by three sisters: Clotho, the spinner, created the thread of life (hence "clothes"). Lachesis, also known as the dispenser of lots, took the measure of that thread, then wove it into the destiny of each man. And Atropos (think atrophy) cut those threads at the time of a man's death. According to Greek mythology, the Fates alone determined a man's destiny; even the Gods could not alter it. The Fates are also credited with creating the alphabet—the very encoded text by which Google prospered. It was said that the Fates cast men's fate through "lots"—runes that each bore a symbol or letter. Each letter reflected an event a person was fated to experience.

10. Google Today, Google Tomorrow

1. I got a firsthand taste of this when my request for a final interview met with a rather bizarre counterproposal from Page. In exchange for sitting down with me, Page wanted the right to review every mention of Google, Page, or Brin in my book, then respond in footnotes. Such a deal would have been nearly impossible to realize, and would have required untold hours of work on Page's part. Page and I negotiated for weeks over his proposal, which communications chief David Krane insisted Page was dead serious about. In the end, Page relented. When we finally met, he apologized for any undue stress his proposal may have caused me, but added that he felt that journalism in general was extremely flawed, and that he was just trying to come up with a way to make it better.
2. Reid was fired—ironically—on Friday, February 13, 2004.
3. Also noteworthy: a 2005 study by the Pew Charitable Trusts noted that the majority of search users cannot distinguish between paid listings and regular search results.
4. Later in the year, Ask Jeeves, the perennial fourth player in search, was purchased by Interactive Corp., an Internet conglomerate that is run by Barry Diller, himself quite a media macher.
5. While Google certainly does have an extraordinary infrastructure, it is not limitless. This fact was proven in early May 2005, when the company

introduced a beta software program it called the Web accelerator. This program leveraged Google's network and bandwith to speed up a user's surfing, in essence by using Google's own servers as proxies for the Internet. The Web accelerator was derided by many Webmasters for various implementation drawbacks, and Google halted the beta distribution, claiming it had hit an internal target for usage. Many believed that Google pulled the program because of the Webmasters' complaints, but a source at the company who is close to the program disputes that fact. "We ran out of bandwidth," the source told me. "It's as simple as that."

11. Perfect Search

1. For more on the future of photography and how digital photographs are becoming searchable, check out Google's Picasa application, or Flickr (now a Yahoo service). For more on the future of video search, try the video search tools from Google, Yahoo, or AOL.
2. Chris Sherman and Gary Price, *The Invisible Web* (Medford, NJ: CyberAge Books, 2001).
3. I am going to resist the urge to digress into a rant on the issue of intellectual property here. However, if you want a good one, read Lawrence Lessig's *Free Culture* (New York: Penguin Press, 2004).
4. It's amazing how fast the search industry is evolving—tools to track search history now exist for nearly every major search engine, all announced in the past year. I now use a tool from Amazon's A9 service to do exactly what I could not do in the summer of 2004.
5. For more, go to www.acm.org/ubiquity/interviews/v5i29_jain.html.
6. www.ftrain.com/google_takes_all.html.
7. I should note that this is perhaps the only digression into the immense field known as enterprise search in this entire book. Why? Two reasons. First, one has to draw lines somewhere—and I decided to focus on consumer-facing search. And second, to be entirely honest, I covered enterprise software for five years at the beginning of my career, and despite how important and lucrative this market will most likely prove for search, it bores the pants off most people. I lack the skills to make it otherwise.
8. Domain-specific vertical search engines in more consumer and commercial domains—such as cars.com or Expedia—are further polluted by the commercial interests of the industries they serve. They could learn a lot from the GlobalSpec approach.
9. They also will create important data mines of user behavior—GlobalSpec has the parametric details of every search ever made against every product

in its database, which is a gold mine for companies who are trying to fathom what the market wants.

10. For good examples, head to Indeed.com (jobs), Oodle.com (listings), and Topix.com (local news).
11. The library of Alexandria was considered by the ancient Greeks to be the apogee of all human wisdom. It burned to the ground in 47 B.C.

Epilogue

1. Though I must admit that in fact I *did* end up buying the epic *Gilgamesh* in print form, after all. Score one for paid search . . . and for the power of print as an archival vestige of search.
2. www.wsu.edu/~dee/MESO/GILG.HTM.

Notice

Index